EXTREME MEASURES

BY THE SAME AUTHOR

Fly: An Experimental Life

EXTREME MEASURES

THE DARK VISIONS AND BRIGHT IDEAS OF FRANCIS GALTON

MARTIN BROOKES

BLOOMSBURY

Published by Bloomsbury Publishing, New York and London
Distributed to the trade by Holtzbrinck Publishers

All papers used by Bloomsbury Publishing are natural,
recyclable products made from wood grown in well-managed forests.
The manufacturing processes conform to the environmental
regulations of the country of origin.

Library of Congress Cataloging-in-Publication Data

Brookes, Martin, 1967–
Extreme measures : the dark visions and bright ideas of Francis Galton / by
Martin Brookes.
p. cm.
ISBN 1–58234–481–7 (hc)
1. Galton, Francis, Sir, 1822–1911. 2. Geneticists—England—Biography.
3. Eugenics. [DNLM: 1. Galton, Francis, Sir, 1822–1911. 2. Eugenics—Biography.
3. Genetics—Biography. WZ 100 G1815BB 2004] I. Title.

QH429.2.G35B76 2004
576.5′2′092—dc22
2004007286

First U.S. Edition 2004

1 3 5 7 9 10 8 6 4 2

Typeset by Hewer Text Ltd, Edinburgh
Printed in the United States of America by Quebecor World Fairfield

For my parents
for their nature and their nurture

Contents

Acknowledgements

Extreme thanks to Jennifer Brady for researching, editing, and listening. Thanks also to David Young, Rosemary Davidson, Catherine Blythe, John Lambert, and DD.

Prologue

Dead on Arrival

The train leaves Stratford-on-Avon on a crisp, clear January morning. Outside the window the landscape changes rapidly from grey to green as a classic country scene rolls into view. The soft, marshmallow hills, the well-kept hedgerows framing fields of ruminating cows, the solitary ancient oak, the isolated cottage with its cast iron gate and a stream running through the garden. It's a brief glimpse into an idealised past. When the Norman tower of Claverdon church pokes out from above a ridge I know that I'm almost there.

I don't see a soul on the half-mile walk into the village. The whole place seems deserted. Even the houses are difficult to spot, sitting at the end of long, shingle driveways and camouflaged by expensive landscaping. But at least there's a pub, the Red Lion, for a pint and a bite to eat. There's just one car in the car park. It's a Rolls-Royce.

Inside, the atmosphere is quiet and respectable. There is no sound of a pool table or a jukebox, just the gentle murmur of polite conversation as a few retired folk enjoy a white wine or a half-pint with their lunch. The bar menu suggests that the food is a cut above your average fare, but the elderly couple at the table next to me are not happy. He ordered the lamb, but it appears it was undercooked. So he's turned away from the meal and gone back to his *Daily Telegraph*. The woman sitting opposite, probably his wife, seems irritated. But she doesn't speak. There is an edgy silence hanging over the unfinished plate of plump, pink lamb.

Suddenly she chirps up. 'How's your knee?' He doesn't seem to hear the question and carries on with his paper. He cuts quite a figure with his blue V-neck sweater and red, polka-dot cravat. As he turns to the business pages his wife repeats her question at a volume that the whole pub can hear. 'HOW IS YOUR KNEE?' He dips the paper below his eyes. There is a slight pause, and then a quiet, almost whispered response, 'I don't know, I haven't asked it.'

Claverdon is turning out to be quite a place. I'm feeling much less apprehensive as I head out to the church after lunch. The blue sky above, however, can't hide the fact that it's blowing a gale. Here, in the heart of the Midlands, I must be at least fifty miles from the nearest beach. But with the wind rattling through the trees it sounds like I'm surrounded by breaking waves.

The graveyard has obviously been busy. The place seems to be overflowing with memorials to the dead. Without knowing where to look it's going to be difficult to pick out the one that I want. So it's fortunate that I find a man on the other side of the churchyard who knows all about these things. John Lambert, reader for the parish, is tending to a huge pile of burning holly. The flames are flying off horizontally in the force-eight gale, licking the paint off some old iron railings. Francis Galton's bones, it seems, are the only things keeping warm on this freezing winter's day.

Lambert has conveniently set up his fire right next to Galton's tomb, and I peer through the railings to get a better look. There is a chill in the air that has nothing to do with the weather. It's the fact that Galton's body is lying there beneath the stone. I can't help thinking that if I knocked on this great horizontal slab of a door, he would suddenly burst out and punch me for being such an impertinent fool.

Galton lies alongside his mother and father in an iron-fenced enclosure, isolated from all the other gravestones. Yet however grand this hallowed cage once appeared, it now looks tired and scruffy. The grave seems to have been abandoned by the living.

There are no fresh flowers, just a forest of moss and lichen, creeping over the stone, as if to mask his memory. Lambert tells me that visitors are rare. The only sightseers, he says, are Japanese day-trippers, taking a break from the Shakespeare experience in nearby Stratford-on-Avon.

Lambert, affable and garrulous, seems to know quite a lot of Galton's story, at least those parts of it befitting the ecclesiastical location. We chat outside for a while before he takes me into the church for a look at something else. Fixed to the wall just in front of the altar is a memorial to the man himself:

In memory of Sir Francis Galton F.R.S.
Born 16th Feb. 1822 Died 17th January 1911
Youngest son of Samuel Tertius Galton and
Frances Anne Violetta, his wife, daughter of Erasmus Darwin F.R.S.
Many branches of Science owe much to his labours
but the dominant idea of his life's work was to
measure the influence of heredity on the mental
and physical attributes of mankind.

The words do polite justice to Francis Galton's career but they barely scratch the surface of his story. Galton was a remarkable character whose long life straddled the entire Victorian era. He came into the world in the wake of Wellington and Waterloo, and went out at the dawn of motor cars and aeroplanes. Living at a time when much of science was still in its infancy, he excelled at a vast range of subjects that would be unthinkable today. He was an explorer, geographer, meteorologist, psychologist, anthropologist, biologist, and statistician. He was not only a polymath, but a pioneer who laid the foundations of modern human genetics. He was a man of prodigious energy, who climbed to the top of scientific societies, wielding power and influence wherever he went. He wrote half a dozen books and over 300 scientific articles. And he won a bagful of awards and a knighthood.

Yet for all his achievements, it is the combination of his feats and his flaws that make him such a fascinating figure. His over-active mind was both a benefit and a burden, a supremely effective tool, but as demanding as an addiction. When mountaineers are asked why they risk their lives to scale a precipitous peak they often reply, 'Because it's there'. Galton seemed to apply a similar philosophy towards counting and measuring. Many of his studies were measuring for measuring's sake, the product of an obsessive drive he possessed from childhood. Galton's obsession unwittingly turned him into one of the Victorian era's chief exponents of the scientific folly. Experiments in tea-making, for instance, were a particular favourite. Galton could not just accept what came out of the pot. Instead, he had to devise complex mathematical equations to work out the best way of making a good brew, based on such crucial considerations as the temperature of the water and the time taken for stewing. Slicing a cake was also seen as a mathematical challenge, and his solution, 'Cutting a round cake on scientific principles', was no doubt eagerly devoured by readers of a 1905 issue of *Nature*.

Galton was a talented inventor. Many of his scientific measurements were obtained using apparatus of his own design. But here again, things sometimes took a peculiar turn. Fearful, for example, that his mind was in perennial danger of overheating, he added a hinged lid to a top hat to provide the necessary ventilation. The lid was raised and lowered by means of a rubber bulb that dangled stylishly from the brim. Other wayward inventions included a pair of spectacles designed for underwater reading, and a bicycle 'speedometer' that consisted of nothing more than a sandglass which the rider was supposed to hold while counting the revolutions of the wheel. It never caught on.

Galton possessed a potent mix of wisdom and whimsy. But other aspects of his character were less appealing. An immense snob, perennially preoccupied with distinctions of race, class, and social status, he was routinely dismissive of those he considered beneath him – women, black people, and the poor. He could be charming and tolerant to family and friends, but heartless and cruel to others.

His cheerful, witty exterior concealed an exceptionally private man haunted by mental illness. His diaries – each less than two inches square – chronicled an outline of his life in minuscule, barely legible handwriting. His most intimate thoughts were recorded in code and then destroyed. So we are left with notes, articles, letters, an autobiography, and the thoughts of others to piece together a picture of this extraordinary individual.

Born in Birmingham in 1822, Galton was the youngest of seven children. His family had made a fortune from manufacturing and banking, and was renowned for its rich scientific heritage. His grandfathers, Samuel Galton and Erasmus Darwin, had both belonged to Birmingham's influential Lunar Society, which brought together some of the greatest figures of the industrial revolution. He also had a cousin who would make his own mark on science. His name was Charles Darwin.

Galton was a shy, independent child whose exceptional abilities were obvious from an early age. Family expectations weighed heavily on his shoulders, and he was pushed into a medical career to which he was eminently ill-suited. But the death of his father in 1844 enabled Galton to abandon his academic studies and turn, instead, to the extravagant life of an English country gentleman.

Ultimately these wilderness years left him unfulfilled. Searching for some higher purpose he re-invented himself once again, this time as an African explorer. Galton showed himself to be remarkably adept at dealing with both the physical and psychological challenges of exploration. Heading out into the hostile environment of the Namibian interior, he re-emerged two years later to international acclaim.

Satiated by adventure, Galton returned to England, seemingly intent on a life of domestic bliss. In 1853, he married Louisa Butler, the daughter of a suitably professional family. The marriage, however, was anything but orthodox. By his own account, marriage was about gaining access to the intellectuals in his wife's family, as much as to Louisa herself. Significantly, the union was a childless one.

By the late 1850s Galton had reinvented himself as a meteorologist. Assimilating weather data from all over Europe, he came up with novel ways of representing the weather in pictorial form. His contoured weather maps set the pattern for those that we see today. He also made an important breakthrough in our basic understanding of how the weather works. Meteorologists had already discovered the cyclonic weather system; Galton discovered its close relative, the anti-cyclone.

But not even Galton could have predicted the great storm of 1859, when Darwin's *The Origin of Species* threw apes and man onto the same evolutionary stage. Like so many others of his generation, Galton was completely blown away by his cousin's evolutionary argument. With God now gone from his life, he felt reborn and revitalised by an entirely new kind of knowledge.

The impact of Darwin's evolutionary ideas extended far beyond the boundaries of biology. The 'survival of the fittest' and the 'struggle for existence' were powerful and compelling metaphors that found social resonance in an increasingly mechanised, industrial Britain. Social Darwinism crept into all walks of life, as Europe became obsessed with notions of individual, racial, and national struggle. You can find evidence of this Darwinian language in the novels of Émile Zola, the poems of Rudyard Kipling, the economics of Herbert Spencer, and the eugenics of Francis Galton.

To Galton, evolution by natural selection was more than just a description of biological history. It was also a prescription for human progress, a natural law that lit the road to a new nation of biological excellence, exceptional in both body and mind. He was convinced that selective breeding was the way forward for society. If it could work with pigs, cattle, and maize, then why not with people?

In the 1860s few people shared Galton's enthusiasm for his vision of a nation of supermen. In some ways, you couldn't really blame them. His Utopian dream sounded like an exclusive, invitation-only rutting club for his well-heeled friends and heroes. By the turn of the century, however, British support for eugenics had grown amid fears that the country was in decline. It was the desire for national improvement that prompted such disparate political voices as John Maynard Keynes, H.G. Wells, and George Bernard Shaw to lend a sympathetic ear to Galton's proposals.

Galton's recipe for human improvement became the stimulus for his pioneering work in heredity, psychology, anthropology, and statistics. He recognised that to better the breed he first had to prove that differences in intellect, energy, and athleticism were all hereditary qualities. Since heredity was pretty much a blank on the map of human knowledge, this left him with some serious work to do. But Galton seemed undaunted by the prospect and was carried along on a wave of near-religious fervour. Before anybody knew anything about genes or chromosomes he set out to uncover the origins of our mental and physical faculties, quickly establishing a reputation for himself as the most thoroughgoing determinist of his, or any other, generation.

Age did little to slow his momentum, and he continued to forage far and wide for his scientific fixes. When he wasn't trying to find novel ways of measuring intelligence or the shape of the human nose, he was devising new statistical techniques to analyse his ever growing mountains of data. Nothing, it seemed, could escape his voracious scientific appetite. Studies in visual perception, essays in extraterrestrial communication, and lessons in self-induced psychosis were all part and parcel of an average day in the life of Francis Galton.

Galton's Victorian life still echoes in our twenty-first century world. Take a trip to University College London, for instance, and you will find the Galton Laboratory, where human genetics was born. When a new medical study reports a correlation between something we like and something that is bad for us, we are witnesses to his ground-breaking steps in statistics. Turn on your television, and you will see evidence of his restless mind in the contours of the TV weather chart. And when a criminal investigator dusts a door frame at the scene of a crime for any tell-tale clues, he is tipping his metaphorical hat to Galton's pioneering work on human fingerprints.

History, however, has not been kind to Galton, and it is easy to see why. Looking back from the safe shores of a new millennium, his philosophy of human progress looks dumb and sinister when set against the excesses of twentieth-century racism. Thirty years after his death, eugenics mutated into the Nazi nightmare of the 'Final Solution'. If Galton wasn't turning in his grave then he certainly should have been.

The appalling events of the Second World War extinguished enthusiasm for eugenics. Yet eugenics arguments have not entirely gone away. Today, they live on in the consultancy rooms of genetic counsellors, and in the offices of life-assurance companies, pharmaceutical organisations, and government select committees. We're still placing biological value on people's lives and futures. It's just that today, the hard, Victorian edge of Galton's eugenic vision has been replaced by the smooth veneer of medical respectability.

Galton, of course, was always destined to live in the shadow of his more illustrious cousin. Take a stroll around the University College London campus and the contrasting fortunes of these two men are all too apparent. Both Darwin and Galton have buildings named after them. But whereas Darwin gets a grand, neo-classical edifice on busy Gower Street, Galton gets a 1960s carbuncle on a gloomy alley behind Euston Station.

Few scientists can hope to match Darwin. But it never stopped Galton from trying. In the end, his scientific life was a complex, messy affair that often seemed to branch off in all directions at once. It was a scientific journey by the scenic route, with frequent excursions to the fringes of madness.

1

Lunar Orbit

Everything we possess at our birth is a heritage from our ancestors.
Francis Galton

On 14 July 1791 tensions were running high in the city of Birmingham. That afternoon an assortment of political radicals and reformers had gathered at the Royal Hotel in the centre of town for a dinner to commemorate the second anniversary of the French revolution. In the two years since the fall of the French monarchy Britain had seen its own constitution dragged into the spotlight. Liberty was a contagious idea and its symptoms were manifest in the many outspoken voices now calling for religious and political reform. For kings and rulers these were nervous, unsettling times. The population had not been this politicised since the start of the Civil War. Prime Minister William Pitt and his Tory oligarchy didn't need opinion polls to spell out the problem; the evidence spoke for itself. Edmund Burke's censorious *Reflections on the Revolution in France* had sold an impressive 19,000 copies. But Thomas Paine's radical reply, *The Rights of Man*, sold ten times as many.

The meeting of reformers at the Royal Hotel was no call to arms. It was a peaceable and good-natured event with an even, diplomatic spread of tributes that included toasts to both the 'King and Constitution' and the 'National Assembly of the Patriots of

France'. But while the event was hardly treasonable, it was certainly ill-timed.

That summer Birmingham was in the grip of social uncertainty. A rise in local taxes had been compounded by a slump in manufacturing, and a restless workforce was demanding answers. The Royal Hotel provided a convenient focus for their complaints. The meeting had attracted many of Birmingham's most influential men, the captains of industry who had turned the town into one of the most prosperous mercantile centres in the country.

By late afternoon an angry crowd had gathered outside the hotel and the atmosphere was lawless and vengeful. Ringleaders loyal to the Crown had stirred up a malevolent brew, turning local grievances into a full-blown confrontation over constitutional reform. The protest was no longer just about jobs but Britain's entire political future. Loud chants in support of the Church and King echoed around the streets. Stones were thrown and windows broken.

To many among the crowd Joseph Priestley was the ultimate symbol of political dissent. For years this renowned chemist and Unitarian minister had been preaching his non-conformist message. Priestley was committed to political freedom and the separation of church from state. He was also extremely sympathetic to revolutionary France.

Priestley had been due to attend the meeting, but concerns over safety had forced him to pull out at the last minute. It turned out to be only a temporary reprieve. If Priestley wouldn't come to the mob then the mob was happy to go to him. As the crowd began to move *en masse* down Bull Street, local magistrates staggered out of the Swan Tavern to cheer them on their way with sly, approving smiles and a toast to 'Church and King'.

The meeting houses, where Priestley gave his religious and political sermons, were first on the hit list. Here the vandalism began in earnest. Pews were pulled out and pulpits smashed. Cushions were torn and thrown and then ignited as both buildings went up in flames. But the conflagration did little to appease the crowd. With their anger now fortified by alcohol, hundreds marched on towards Birmingham's southern suburbs, and Priestley's house at Fair Hill.

Priestley was at home playing backgammon with his wife Mary when news came in of the approaching hordes. They managed to slip away just before the first violent wave arrived. His house was ransacked. His laboratory was reduced to rubble. From his vantage point on a nearby hill, the founding father of oxygen chemistry could only watch as his lifetime's work went up in smoke.

The mob's wrath that night remained unabated. Anyone who shared Priestly's political sympathies was considered fair game, and over the next few days the homes of dozens of known dissenters were systematically sought out and destroyed.

Priestley and his wife were forced to keep on the move, shifting from one house to the next in order to avoid detection. The couple had no shortage of friends who were prepared to risk their own lives to help them. One such ally was Samuel Galton, Francis Galton's paternal grandfather.

Samuel Galton's association with Priestley went back to his childhood. In 1768, at the age of fifteen, he had been sent to Priestley's Warrington Academy in Cheshire. Under Priestley's guidance, the school had established a learning environment that mimicked his own thirst for knowledge. He introduced science into the curriculum, a radical move for the day, and constantly encouraged his pupils to think for themselves and question the world around them. It was that inquiring philosophy that would eventually bring the two men together in Birmingham fifteen years later.

At the tail end of the eighteenth century Birmingham was a great place for any aspiring entrepreneur, a place where scientific thinking catalysed business brains into shaping the industrial and commercial landscape of Britain. Enlightenment think-tanks flourished in this climate of technological optimism, and few were more illustrious than the Midlands-based Lunar Society. Meeting each month on the Monday nearest the full moon, the society brought together a loose collection of scientists, engineers, philosophers and industrialists. Among its members were the steam pioneers Matthew Boulton and James Watt; William Withering, the botanist who isolated the heart drug digitalis from foxgloves; and the innovative potter, Josiah Wedgwood. Joseph Priestley was also a member, and so was Samuel Galton.

The Galtons were a Quaker family who had moved to Birmingham from Bristol in the 1740s. Samuel's father was a self-made man who had started out from modest beginnings and built a business empire in property and gun manufacturing. Samuel Galton took over the gun factory in 1774, when he was only twenty-one. Like his father, he became an astute businessman, and supplemented his income from the factory with shrewd investments in livestock and agriculture, property and canal development.

Francis Galton described his grandfather as 'a scientific and statistical man of business', who loved 'to arrange all kinds of data in parallel lines of corresponding lengths, and frequently using colour for distinction'. There was no doubting Samuel Galton's soft spot for statistics. He used elaborate charts and tables to describe every aspect of his income and household expenditure, and he harvested facts and figures about horses, canals, building materials, and anything else that aroused his interest. This numerical predilection would be shared by, and come to shape the life of, his future grandson.

While business interests dominated Samuel Galton's working life they could not obscure his passion for science. At the age of twenty-one he was forming his own scientific library. He read French naturalist Georges-Louis Leclerc Buffon's seminal and voluminous masterpiece *Histoire naturelle*. He attended public lectures on optics and the nature of gases. And he acquired a microscope, a reflecting telescope and an assortment of electrical machines to allow him to get deeper into the natural world.

As he absorbed the scientific world around him, Samuel Galton turned from science student to active exponent. Using spinning,

Samuel Galton (1753–1832), Francis Galton's paternal grandfather

circular cards with different-coloured segments he pioneered the idea of primary colours, research which served to cement his reputation. He was elected as a Fellow of the Royal Society in 1785.

Beyond his contribution to the understanding of colour, Samuel Galton published relatively little. His fascination with ornithology found expression in a three-volume work on the natural history of birds, and he also wrote a short paper on canal levels. Compared to some of his Lunar Society friends his scientific achievements look modest. Yet his commitment to scientific endeavour was unquestionable, and he was extremely generous in the support of others. Joseph Priestley's research, in particular, owed much to Samuel Galton's financial backing.

In 1777 Samuel Galton married Lucy Barclay, a great-granddaughter of the Quaker apologist Robert Barclay. The Barclays were noted for their robust constitutions. Lucy was certainly a sturdy and formidable-looking woman and her half-brother, Captain Robert Barclay-Allardyce, was fabled for his amazing feats of strength. He literally walked his way into fame when he covered an astonishing 1,000 miles in 1,000 hours. It was from the Barclay side of the family, Francis Galton insisted, that he received his 'rather unusual power of enduring physical fatigue without harmful results'.

Samuel and Lucy had ten children in all, six of whom lived well into their old age. Mary Anne, born in 1778, was the eldest. She remembered a happy, prosperous, and, above all, scholarly childhood. Her mother used to sit her on her knee and read Buffon's *Histoire naturelle* to her. The conscientious parents were ambitious for their children and took great to pains to encourage an interest in all aspects of science and philosophy. Childhood pets, for instance, were not just animals to be pampered and stroked; they were objects to be assigned to the Linnean system of classification based on the structure of their jaws and the details of their teeth.

The ambience of the family home was in tune with the kind of illustrious circles in which the Galtons were moving. Samuel and Lucy played host to numerous Lunar Society meetings. Mary Anne recalled the atmosphere on one occasion in 1788. Chief industrialist Matthew Boulton had just returned from Paris and the whiff of revolution was in the air.

It was wonderful to me to see Dr Priestley, Dr Withering, Mr Watt, Mr Boulton himself and Mr Keir, manifest the most intense interest, each according to his prevailing characteristics, as they almost hung upon his words . . . My ears

Captain Robert Barclay-Allardyce (1779–1854), Francis Galton's great-uncle

caught the words, 'Marie Antoinette', 'The Cardinal de Rohan',
'diamond necklace', 'famine', 'discontent among the people',
'sullen silence instead of shouts of "Vive le Roi"'.

That same year a special guest came round for tea. He was the
family physician and one of the original brains behind the Lunar
Society. The Galtons had laid on an extravagant spread of tropical
fruits, cakes, pastries, sweets, clotted cream, and the finest Stilton
cheese. Having fixed his eyes on the feast, the ravenous guest
shifted his corpulent, twenty-stone frame into position and proceeded to eat the lot.

An excessive appetite was just one small part of Erasmus
Darwin's larger-than-life character. Darwin made his name as a
medical man, but he was truly a man of ideas and liked nothing
better than to wallow in the possibilities of his enlightened and
industrial age. As a poet, naturalist, and inventor he was certainly a
model for one of his future grandsons, Francis Galton.

Darwin's reputation as a physician was so great that the royal
court had tried to secure his services to treat the maladies of
George III. But London held no attraction for Darwin and, besides, he enjoyed his itinerant existence. Touring the country in
his bone-shaking stagecoach he was something of a Robin Hood
of the medical world, charging the Galtons and other prominent
Midlands families extortionate amounts, while asking nothing
from the poor. But while medicine earned him a living it also kept
him away from the Lunar Society. 'I am sorry the infernal
Divinities, who visit mankind with diseases, and are therefore
at perpetual war with Doctors, should have prevented my seeing
all your great Men at Soho today', he wrote to Matthew Boulton
in 1778.

Lord! what inventions, what wit, what rhetoric, metaphysical,
mechanical, and pyrotechnical, will be on the wing, bandy'd
like a shuttlecock from one to another of your troop of
philosophers! while poor I, I by myself I, imprizon'd in a

post chaise, am jogged and jostled, and bump'd, and bruised along the King's high road to make war upon a pox or a fever!

Erasmus Darwin (1731–1802), Francis Galton's maternal grandfather

Early on in his medical career Erasmus Darwin had lived in the small cathedral city of Lichfield, thirty miles or so from Birmingham. There he met, fell in love with and married a neighbour, seventeen-year-old Mary Howard.

Erasmus and Mary Darwin's first child, Charles, followed in his father's footsteps and studied medicine, but died before he ever got the chance to practice. When Robert, the youngest son of Erasmus and Mary, became a father, he named one of his sons for his late brother. This Charles Darwin would also start life as a medical man, but it was his theory of evolution that would immortalise the family name.

Five children came from Erasmus and Mary's marriage. There might have been more had Mary not died so young. In her late twenties she began complaining of serious abdominal pains. Erasmus prescribed opium, which worked well for a while, but Mary turned increasingly to alcohol to stem the pain, first wine and then spirits in ever larger quantities. She died from cirrhosis of the liver, terrified and delirious, at the age of thirty.

Devastated by the death of his wife, Erasmus Darwin took on a nanny, Mary Parker, to look after his children. Within a year the eighteen-year-old was pregnant with Erasmus's child. The couple lived together throughout the early 1770s, although they never married and the relationship petered out after the birth of their second child. By 1775 Erasmus was courting the affections of another younger woman. This time, however, the circumstances were quite different. Elizabeth Pole, the latest object of his desire, was already married.

Elizabeth had the kind of looks, intelligence and wit that won many admirers, with a tall, slim figure, dark hair, and a wide benevolent smile. She was also lively, affectionate, and adventurous. In 1769, in her early twenties, she had married the dashing Colonel Edward Sacheverel Pole, a man almost thirty years her senior, and a celebrated British warrior and veteran of the Seven Years' War.

Erasmus Darwin first met Elizabeth in 1775, when he took over as the new family physician. Immediately he fell under her spell. Despite the obstacles of her heroic husband and three young children, Darwin, always the optimist, was undaunted. With no sword to wield he put all his faith in his pen, pouring out his heart in a deluge of florid love poems.

In 1780 Darwin's romantic ambitions were given a boost by the death of Colonel Pole. But when Elizabeth was quizzed on the possibility of hitching up with Darwin, she sounded less than enthusiastic: 'He is not very fond of churches, I believe, and if he would go there for my sake, I shall scarcely follow him. He is too old for me.' She married Darwin six months later.

Erasmus Darwin was almost fifty when he wed Elizabeth, but he showed no signs of slowing down. On the contrary, the apparent success of his love poems gave him the confidence to embark on more ambitious projects. *The Loves of the Plants*, published in 1789, was ostensibly a book about botanic sex, told in thousands of rhyming couplets. But in Darwin's capable hands plants became a medium through which he explored the interacting forces in his scientific and industrial age. The critics loved it, and Darwin caught the attention of young poets, like Wordsworth and Coleridge. But the book's overwrought romanticism soon became dated, and Francis Galton described the poetry as 'intolerable'.

Darwin persevered with other literary projects. His four-volume medical epic, *Zoonomia*, was more conventional and all the better for it. The book was a distillation of decades of medical experience and covered all aspects of mind and body. Darwin's medical vision was typically expansive and included some serious doses of speculation. One chapter, 'On Generation', dared to dream the impossible; a biological world without a Creator, driven by its own internal engine:

Would it be too bold to imagine that all warm blooded animals have arisen from one living filament, which THE GREAT FIRST CAUSE endowed with animality, with the power of acquiring new parts, attended with new propensities, directed by irritations, sensations, volitions and associations; and thus possessing the faculty of continuing to improve by its own inherent activity, and of delivering down those improvements by generation to its posterity, world without end!

Darwin and Elizabeth had seven children. Violetta, their first child, would become Francis Galton's mother. But it was Francis, Violetta's younger brother, who was the prototype for his nephew and future namesake. After an abortive medical career Francis Darwin dedicated himself to a life of adventure. In 1808, aged twenty-two, he set off with four others on a two-year tour of the Mediterranean and the Middle East. The trip was blighted by war, pirates and disease. Of the five members of the party who originally set out only Francis Darwin returned alive. Theodore Galton, son of Samuel and Lucy, was among the four who died.

Friendships fostered by the Lunar Society embraced entire families. The Society bonded Darwins with Galtons, Watts with Wedgewoods. It was a conduit through which the sons and daughters of the industrial age could meet and let off some steam. When Samuel Tertius Galton fell in love with Violetta Darwin, it turned the cog a notch nearer to the birth of Francis Galton.

In 1804 Samuel Galton wound up the gun business and the factory was turned into a bank, with his eldest son, Samuel Tertius, in charge. Samuel Tertius was a genial and conscientious man who inherited all the wealth, if not the energy and enterprising instincts of his father. His dedication to his duties at work left him with little opportunity for outside interests. He did, however, find time to engage in one small, but significant piece of economic research. His snappily titled article, 'A Chart Exhibiting the Relation between the Amount of Bank of England Notes in Circulation, the Rate of Foreign Exchanges, and the Prices of Gold and Silver Bullion and of Wheat, accompanied with Explanatory Observations' was interesting for several reasons. Not only did it display the family fondness for figures, it also introduced a statistical idea that his son Francis Galton would later make his own. Samuel Tertius used graphics and charts to illustrate the strength with which two variables were associated with one another. Sixty years later, his son worked out a way of assigning a numerical value to this association, and called it the correlation coefficient.

In 1807 Samuel Tertius married Violetta Darwin. Violetta was less conventional than her husband. Sharing much of her mother's good looks and personality, she was effervescent, adventurous and artistic, the perfect foil for her conservative husband.

Violetta Darwin (1783–1874) and Samuel Tertius Galton (1783–1844)

The newlyweds moved into a house in Ladywood, in Birmingham's rural western suburbs, and wasted no time in starting a family. Elizabeth, their first child, was born in 1808. Lucy, Adèle, and Emma followed in swift succession. Health problems plagued Adèle and Lucy from an early age. Adèle was born with a curvature of the spine that forced her to spend much of her childhood prostrate on a board, while Lucy suffered from recurrent bouts of rheumatic fever. But both of them fared better than Agnes and Violetta, two more Galton daughters who died in infancy.

These deaths prompted some serious soul-searching. Acting, perhaps, under Violetta's influence, Samuel Tertius decided to sever the Galton family's long-standing connection with the Quakers and join the Church of England. Samuel's belated baptism was preceded by the birth of two more children, Darwin in 1815, and Erasmus a year later. With four daughters and two sons the Galton family seemed complete. Protected from the harsher realities of industrial Britain by their father's considerable wealth, the family settled down to a life of privilege and luxury in Birmingham's rural suburbs. It was, by all accounts, a very happy household. Violetta's outgoing and affectionate nature complemented Samuel Tertius's pragmatic diplomacy, creating a warm and sympathetic environment for their children. There were arguments and disputes, of course, but Samuel Tertius proved himself to be the consummate peacemaker. 'When we children quarrelled,' Elizabeth once recalled, 'and went to my Father or Mother to complain, he used to send one into one corner of the room, and the other into the opposite corner, and at the word of command, each had to rush into the other's arms. This made us laugh and ended the dispute.' Samuel Tertius's skills as an arbitrator were put to use in public life when he was appointed High Bailiff of Birmingham in 1814.

In 1820 the family moved to a new home, The Larches, in the fashionable country suburb of Sparkbrook, a couple of miles to the south of Birmingham city centre. The property was only a stone's throw from the spot where Joseph Priestley's house had stood, before it was destroyed in the Birmingham riots of 1791. Priestley was long since gone. He had emigrated to the United States shortly after his house burnt down, and died there in 1804. But Lunar links persisted in the area.

At one time The Larches had belonged to the botanist and Lunar Society member William Withering. He bought the house, with its ten acres of land, in 1799, but his tenure had been a short one. Withering moved in on 28 September and died a week later. The house was looked after by his son for the next twenty years until the Galtons moved in, renting the property for £220 a year.

The Larches took its name from the two towering trees that stood guard to the left and right of this classical Georgian country manor. The house was generously proportioned, three storeys high, and five windows across. To the rear, stables, barns and coach houses opened onto acres of wide open fields, offering the perfect playground for the children and their ponies.

A year after moving into their new home Violetta discovered that she was pregnant again. She was thirty-eight years old and it had been six years since the birth of Darwin, her last child. The pregnancy caused great excitement within the family and everyone eagerly anticipated the new arrival. On 16 February 1822, their wait came to an end and baby Francis was born. Sixty years later Emma recalled the momentous day:

> It seems but the other day that Mrs Ryland had called with her 4 horses, and walked in the garden by my mother's garden chair. [Aunt] Booth dined at our house, and in the evening you were born about 9 o'clock. And the importance Darwin, Erasmus and myself thought of the Dudson carriage and pompous coachman coming early on the following morning (Sunday) to take us to spend the day at Dudson [grandfather Samuel Galton's house]. And we worried the servants by every now and then standing on a chair to make us high enough to reach the call-tube in the Library to inform them: 'Mama had a Baby, and it was a *Boy*!'

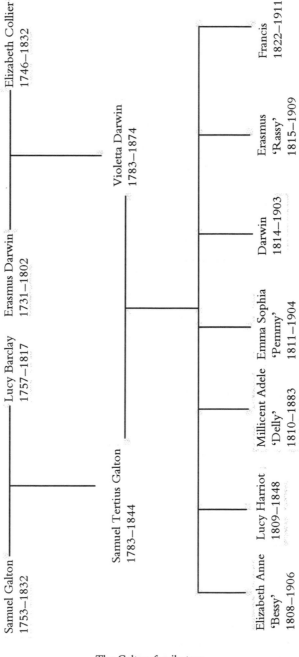

The Galton family tree

2

Boy Wonder

The horizon of a child is very narrow and his sky very near. His father is the supreme of beings.

Francis Galton

The pampering began in earnest. Francis was the centre of attention and the four adolescent sisters quarrelled among themselves for the honour of holding him in their arms. 'He was the pet of us all,' Elizabeth remembered, 'and my mother was obliged to hang up her watch, that each sister might nurse the child for a quarter of an hour and then give him up to the next.'

Twelve-year-old Adèle harboured thoughts of a more substantial involvement in her new brother's upbringing. Her spinal condition had confined her to a couch for most of her waking life, often leaving her isolated from the activities of the rest of the family. The new baby offered an escape from the boredom of this lonely existence. She wanted to be Francis's guardian, his mentor, and his guide, and she pleaded with her mother for the chance. Violetta agreed, and at the age of five months Francis had established his place on the couch by her side.

It was a challenge that Adèle took to with relish. Her bedroom became Galton's nursery, and she invested all her time, energy and ambition in his fate. His education was her number-one priority, and she kept up a ferocious pace, cramming as much information into his infant brain as it could take. 'Her idea of education at that time', Galton recalled, 'was to teach the Bible as a verbally inspired book, to cultivate memory, to make me learn the merest rudiments of Latin, and above all a great deal of English verse. This she did effectively, and the result was that she believed, and succeeded in making others believe, that I was a sort of infant prodigy.' At twelve months Galton could recognise all the capital letters; at eighteen months he was comfortable with both the English and Greek alphabets – and cried if they were removed from sight; at two and a half he read his first book, *Cobwebs to Catch Flies*; a month later he was signing his own name. By the time he was four, he was more than able to spell out his achievements for himself:

> I am four years old and can read any English book. I can say all the Latin Substantives and Adjectives and active verbs besides 52 lines of Latin poetry. I can cast up any sum in addition and can multiply by 2, 3, 4, 5, 6, 7, 8, 10. I can also say the pence table. I read French a little and I know the Clock.

Galton came to the Greek myths a little earlier than most. At the age of five he was already well acquainted with the works of Homer. The *Iliad* was his favourite, and he was known to burst into tears when he got to the point in the plot where Diomed is wounded by Paris. Early exposure to Homer may have inspired his lifelong fascination with Greek culture and society. Years later he would publicly proclaim the Ancient Greeks as the model to which all future generations should aspire.

While Galton's natural abilities were self-evident, there were also signs that Adèle's punishing schedule was pushing her brother's precocious talents too far. He was his own worst enemy, fiercely ambitious and unwilling to accept his limitations. His appetite for learning was enormous. But every now and again, he would drop hints that all was not harmonious in his mind. There was the time, for instance, when his father was testing him on arithmetic. Galton seemed quieter than usual, a little subdued and his father asked him if he was tired. 'I am not tired of the thing, but of myself,' said the five-year-old boy.

Moments when Galton was doing nothing were rare. If he didn't have his head in books then he was outside, acting out his Greek and Trojan fantasies, galloping around the fields on his horse like a young Achilles. Sometimes the pace would slow and he would immerse himself in the local wildlife. He was an avid insect collector and a keen gardener, and he took great pride in tending his small border of flowers. He was a busy, energetic boy and, for the most part, a very happy one. The Larches, the enchanting country manor where he grew up was, in his own words, 'a paradise for my childhood'.

Yet he was also a shy and solitary child with few close friends. His siblings were all too old to make childhood companions. His sisters, to whom he was always most attached, were turning into young women as he was turning five. Concerned for his well-being, Adèle thought it was time Francis met some children of his own age. In the late summer of 1827 he was marched off to a small school in Birmingham.

School came as a bit of a culture shock. At first he feared that his mother would have him instantly removed on account of the ignorance of his fellow pupils. He was stunned to learn that no one had heard of, let alone read, the *Iliad* and other bedtime classics. Yet once he had recovered from the trauma he settled down quickly to a relatively normal school life. He was made head boy and, according to his mother, his 'open candid disposition with great good nature and kindness to those boys younger than himself' made him highly popular among his classmates. His peculiarities remained – he had a tendency to fend off taunts not with fists, but a barrage of classical quotations – but other children learned to accommodate these eccentricities, and even use them to their advantage. On one occasion a fellow pupil approached him for advice on a letter he was writing to his mother. The boy's father, caught up in a political coup in Madeira, was in danger of being shot and the boy was struggling to find the

appropriate words. Thinking for a second, Galton proffered this
fitting couplet from Walter Scott:

> And if I live to be a man
> My Father's death revenged shall be.

Spouting verse became something of a Galton trademark in his
early years. Out riding with his brother Darwin one day, he was
thrown from his horse into a muddy ditch. As Darwin dragged him
out by his feet, Galton was busy quoting from Samuel Butler's
Hudibras:

> I am not now in Fortune's power
> He that is down can fall no lower.

And when the family servants were trying to round up an unruly
flock of geese, he plunged into the maelstrom and emerged,
moments later, with a gander held aloft, muttering lines from
Chevy Chase:

> Thou art the most courageous knight
> That ever I did see.

As a middle-aged man Galton remembered these events with
embarrassment. He feared that he had been an 'intolerable prig',
and questioned family and friends on the subject. They assured
him that he 'was not at all a prig, but seemed to "spout" for pure
enjoyment and without any affectation'. Either way the spouting
didn't last long. His father had big plans for his youngest son.
At the age of eight, he was sent away to a boarding school in
Boulogne.

Samuel Tertius had never wanted to spend his whole working
life in a bank. But the job had consumed his dreams and aspirations and he became, in the words of his son, 'a careful man of
business, on whose shoulders the work of the Bank chiefly rested

in troublous times'. He seemed to divert his own unfulfilled ambitions into plans for the future of his youngest son. The move to France, he argued, was the perfect way for Francis to improve his French accent.

On 3 September 1830 father and son left Birmingham for London, where they caught a steamer bound for Calais. From there it was just a short trip down the coast to Boulogne. Samuel Tertius hung around for a week, 'to assure himself of the dear child's perfect happiness'. And when it was finally time to say goodbye, his son 'did not shed a tear, or seem at all uncomfortable at parting with his father, but to the last repeated how happy and comfortable he was, and how kind Mrs Neive, the housekeeper, and everybody was to him'. Galton would not see his father, or any other member of his family, for nine long months.

He hated the school. French speaking was compulsory and enforced by a strict system of discipline. Any child caught uttering an English word had a metal label pinned to his chest. Accumulate enough of these labels and it was time to bend over for a beating: 'The birchings were frequent and performed in a long room parallel to, and separated from, the schoolroom by large ill-fitting doors, through which each squeal of the victim was heard with hushed breaths.'

Every day the children were taken for a walk, marching in pairs in a long column around the town. Sometimes in the summer they were allowed to go down to the rocky shoreline to swim. These occasions were a great opportunity, Galton remembered, for 'inspecting with awe the marks of recent birchings, which were reckoned as glorious scars'. But the trips also offered precious moments of pleasure. He recalled with glee how he and the other children used to save their bread from breakfast. Then, 'after having gathered mussels, we spread their delicious contents on it to eat'.

Galton tried to remain stoical about his school life in letters to his family, but by late November, and only two months into his stay, his longing for the comforts and company of home was transparent: 'I hope you will come over soon here for I should like to see you, and to go out with you, for I miss dear Papa's green-gages, which he used to give me when he was here.' His appeals fell on deaf ears and he spent both the Christmas and Easter holidays alone in Boulogne.

He returned to England in the summer of 1831 for a family holiday in Worthing, but he was back in Boulogne by September. This period of French isolation proceeded pretty much like the first, with his parents failing to show up again for the Christmas and Easter vacations. Throughout this period Galton had to rely on letters as his sole means of communication with the family. And in 1832 the news was mostly bad. In February he learned about the death of his grandmother Elizabeth Darwin. A few months later, Samuel Galton, his grandfather, also passed away.

Beyond the tears of bereavement, Samuel Tertius and the rest of the Galton family could look forward to a significant inheritance. At the time of his death, Samuel Galton had assets totalling about £300,000. Samuel Tertius was already an affluent man, but his father's legacy took his wealth to new heights, and assured financial independence for his entire family.

Samuel Galton had far outlived his Lunar Society friends. Erasmus Darwin, Matthew Boulton, James Watt, Josiah Wedgwood, Joseph Priestley, and William Withering were all long dead. Gone too was the unbridled optimism that had allowed the Lunar Society to flourish in the first place. British society had changed. Years of bloody war with Napoleon's France had made a population numb to the thrill of discovery. Traditional religion had got back on its feet, ushering in a new era of conservatism. Mary Shelley's *Frankenstein* had turned science into sorcery. Luddites were left out in the cold as the machines roared on, but Enlightenment idealism could now look naive and hollow when set against the social and economic realities of the industrial age.

The death of two grandparents was not the only change to occur while Galton was away in France. His father had retired from the Galton Bank in 1831 and relocated to the small spa town of

Leamington, where he continued to work as a magistrate. In addition, he had bought a quiet country retreat in the Warwickshire village of Claverdon, a bucolic antidote to his worsening asthma. Both of Galton's brothers had left home and settled down to the life of country squires, while sister Lucy had married and was living in Smethwick, west of Birmingham.

Galton, too, was about to feel the changes. In 1832 his father decided to abandon the French experiment. With the nightmare behind him, Galton found a new school and a new lease of life. He attended a small private boarding school run by the Reverend Mr Atwood, a man who, in Galton's words, 'showed so much sympathy with boyish tastes and aspirations that I began to develop freely'. Not only was there some science to augment the customary classics and theology, there was also carpentry, cricket, and archery to fill the gaps between formal lessons.

In this new, enlightened environment, a more contented Galton found himself making friends with his fellow pupils. Chief among these were the Boulton brothers, Hugh William and Matthew P. Watt Boulton, grandsons of Lunar Society original, Matthew Boulton. Galton was in awe of the two boys, especially Matthew, who became a lifelong friend and influence. In his autobiography Galton explained why he was such an 'object of reverence':

I have known few or any who seemed to me his natural superiors in breadth and penetration of intellect . . . His artistic sense was of a high and classical order. His ideal, like that of Goethe, was a uniform culture of all the higher faculties. There was nothing ignoble in his nature. Whenever I talked with him about my own occasional annoyances, they seemed to become petty through his broad way of looking at things, I may almost say under the mere influence of his presence.

Outside of school Galton was growing up fast. Summer holidays with the family to Aberystwyth, Weymouth and Worthing saw the emergence of Galton the gunslinger, out to right wrongs on seagulls, wagtails, and swallows. When he wasn't shooting the birds he was raiding their nests for his egg collection. But these magpie antics could sometimes get him into trouble. On a visit to sister Lucy's house in Smethwick, he spied a bird's nest on a branch overhanging a canal at the bottom of the garden. Climbing up to reach the nest he slipped. The top half of his body went head first into the water, but his legs somehow got left behind in the branches. Suspended, Houdini style, upside down in the canal, he struggled to detatch himself and, like a true showman, left his escape until the last minute.

Galton was having such a good time at Reverend Atwood's school that it was only proper and right he should be removed from it. In January 1835, a few days short of his thirteenth birthday, he was marched off to King Edward's grammar school in the centre of Birmingham. Within a few weeks, however, there was an epidemic of scarlet fever at the school and he was back at home in bed. Galton survived the attack but his friend Johnny Booth did not.

Normal activities resumed after Easter when Galton returned to board at the house of Dr Jeune, King Edward's headmaster, in Edgbaston. But the change of scene presented only a mild improvement on his recent bedridden misery. Under Dr Jeune's stewardship King Edward's was run on the principles of drill and discipline. Each day Galton and his fellow boarders would walk the suburban mile to school to begin a monotonous round of Latin grammar, Greek verse, and traditional English beatings. It was a cruel and ruthless regime in which Dr Jeune led by example. In one Greek lesson he thrashed eleven pupils in eight minutes for imperfect pronunciation; Galton timed him.

The official end of the school day brought little relief from the discipline. Since Galton was boarding with Dr Jeune, the humiliation continued well into the evening. Extracts from Galton's diary for 1836 offer a snapshot of these hostile times:

Tuesday, Jan. 26	The Dr flogged a chap. The Dr's father was buried.
Friday, Feb. 19	Saw a stuffed cat with 6 legs 4 ears 2 tails and one eye.
Tuesday, March 15	One boy was expelled and another flogged.
Saturday, March 19	Took a walk to Edgbaston park. Earp bought a swing for us, to put which up we had to cut away some shrubs; we expect a row.
Sunday, March 20	The Dr made a tremendous row about the swing and said that it should be taken down.
Monday, March 21	The swing was taken down. We set up some leaping posts.
Saturday, March 26	Bought a cat's gallows. Got caned.
Monday, March 28	Got caned.
Good Friday, April 1	We were made to fast, but we went over to the grubshop and got plenty.
Monday, April 18	I knocked a fellow down for throwing a brick at me.
Tuesday, April 19	I thrashed a snob for throwing stones.
Sunday, May 8	A fellow gave me a thrashing in the street.

It was all fairly typical adolescent warfare. If nothing else it showed how much he had changed from the peculiar creature of his infancy. But he was extremely unhappy again. Bored with the classics and weary of the punishments, he yearned to escape from King Edward's. Long letters home to his family were barely concealed pleas for rescue:

I have not been able to write on account of the hard work and many impositions I have lately had – 30 one day and 10 pages of Gk. grammar to write out, the next 40, and the next 40, so that I have not had the least time. Another boy has left and is believed to be in a consumption. Indeed I never knew such an unhappy and unlucky school as this; 2 more will leave at Christmas, and I would give anything if I could leave it too.

Fortunately, salvation lay just around the corner. By late 1837 his father was making plans to remove him from King Edward's. There was to be an entirely new direction to his education. Galton's mother Violetta had always wanted Francis to follow in the footsteps of her father, Erasmus Darwin. So that was that. Whether he liked it or not, Galton was going to become a medical man.

Samuel Tertius immediately set about organising his son's enrolment at Birmingham's General Hospital. The fees of £200 a year were not insubstantial and Samuel Tertius seemed especially keen that this latest experiment in his son's education should go well. Once business at the hospital was taken care of he wrote to his son, outlining his hopes and ambitions. The letter read like a rallying cry, a point-by-point manifesto for Galton's medical success:

> I really believe, if you turn the opportunities you will have at the Hospital to the best account and avail yourself of the advantages of explanation that my medical friends there will be disposed to give you, if they find you willing to profit by them, that you will begin your medical career very propitiously. You must be careful to avoid low company and not be led astray by any pupils there that may not be equally well disposed – but I have great confidence in your wish to do what is right, and when we meet at your approaching holidays, we will talk over all your plans and arrangements in good earnest and particularly in reference to your masters and studies whilst at the Hospital.

The summer of 1838 proved to be a memorable one, not only for Galton, but the whole of the British nation. In June, he joined his sisters and thousands of others in London for the coronation of Queen Victoria. The city had put on its best summer clothes, with buildings and balconies decked-out in a uniform of crimson cloth. Enormous crowds turned out to watch the young queen, still only nineteen, travel through London to Westminster Abbey and take her ceremonial route to the throne. Sister Emma had managed to use her connections to wangle a ticket inside the Abbey itself, but Galton had to be content with a seat among the hoi polloi in Pall Mall.

A month later he was in Brussels, on a whistle-stop European tour. The trip had been organised by his father, who thought that his son should make the most of his holiday and gain some experience before starting his formal medical training in the autumn. Having heard about a couple of young and knowledge-able medical men who were planning an excursion to Europe, he had arranged for his son to tag along.

The tour was supposed to be a working holiday, combining sightseeing with visits to continental hospitals. Judging by his letters home, Galton was having a great time. 'There is certainly nothing more useful than travelling,' he wrote on his way to the Danube. 'The more you see the more you are convinced of the superiority of England. However nothing can be so admirable as a German or Frenchman who loves his country; it must be a great and genuine patriotism to be able thus to prefer it.' Only in Munich did he sound a sour note when the rigours of long, distance travel finally began to catch up with him: 'I have got one boil and two blisters in such awkward positions that when sitting back I rest upon all three; when bolt upright on two, and when like a heron, I balance myself on one side upon one!!!'

In Vienna the three men decided to drop into a local lunatic asylum. While they were touring one of the female wards, a 'young, buxom, and uncommonly good-looking female lunatic dashed forward with a joyful scream'. She clasped Galton tightly to her bosom with both her arms, calling him 'her long-lost Fritz'. This was undoubtedly a traumatic experience for the diffident sixteen-year-old, and his memory of events was still vivid more than seventy years later. 'In those days', he recounted in his autobiography, 'I was particularly shy and sensitive, and a consciousness of even the least unconventionality made me blush to an absurd degree.'

Galton returned to Britain at the end of September 1838. Two weeks later, he was walking through the doors of Birmingham General Hospital to begin his new life as a medical student. He could be forgiven for feeling a slight sense of trepidation. Hospitals of the 1830s were no place for the squeamish. They were gruesome playhouses for the sick and dying. Each dark, damp, featureless ward offered its own grotesque treatment of human suffering, while down in the dungeons, operating theatres hosted atonal symphonies of saw on bone, knife on flesh, and the pathetic screams of patients. This was still a primitive medical age, ignorant of antiseptics and anaesthesia.

Galton spent his first few weeks in the hospital dispensary, learning how to make pills and medicines. Keen to understand the effects of the compounds he was dispensing, he decided to try all the drugs out for himself, starting at A and continuing in alphabetical order. He safely negotiated his way through the As and Bs – aconite, adder's tongue, aniseed, benzoin, bryony, and burdock – and had almost got to the end of the letter C, when he came up against Croton oil. Used to treat tetanus and mania, Croton oil also had a formidable reputation as the most powerful of all known purgatives. 'I had foolishly believed that two drops of it could have no notable effects', remembered Galton, 'but indeed they had.' The oil purged not only his bowels, but his will to continue the experiment, and drugs D to Z were left untouched.

Within a few weeks he had moved out of the pharmacy to join the real cut and thrust of hospital life. Now he was accompanying surgeons on their morning rounds, attending accident rooms, operating theatres and post-mortem examinations, learning fast how to set broken limbs, treat burns and dress wounds. 'There is an immense deal of work here,' he explained in a hastily written letter to his father. 'It does not come in one long pull but in a series of *jerks* of labour between intervals of rest, like playing a pike with a click reel.' He went on to describe one of his typical working days:

> ½ past 5 p.m. went round all the wards (No joke I assure you) – made up about 15 prescriptions. Awful headache etc. Entered in the Hospital Books records etc. of patients; writing in my case book etc., hard work till 9. Supper. Went round several of the wards again. Accident came in – broken leg, had to assist setting it. ½ past 11, had to read medicine etc. 12, very sleepy indeed, lighted my candle to go to bed. A ring at the Accident Bell; found that it was a tremendous fracture. Was not finished till ½ past 1. Went to bed and in the arms of Porpus. 3 a.m. in the morning: a tremendous knocking at the door; awful compound fracture, kept me up till 5. Went to bed – up again at 7 o'clock. – Rather tiring work on the whole, but very entertaining.

He was deeply affected by the human suffering that he witnessed within the hospital, and always impressed by the dignity of those confronted with their own deaths. One death-bed scene left a lifelong mark on his memory: 'A girl was fast dying of typhus, and I had been instructed to apply a mustard plaster. When I came to her, she was fully sensible, and said in a faint but nicely mannered way, "Please leave me in peace. I know I am dying, and am not suffering." I had not the heart to distress her further.'

Daily exposure to death and disease inevitably engendered a partial immunity to the personal tragedies unfolding in front of him. His first visit to the operating theatre provoked a typical response of horror and revulsion. But this soon gave way to a more detached, scientific, point of view. There was still pity, but the cries of pain coming up from the operating table were now seen as important pieces of information, data points in some vast and terrible experiment into human suffering. Galton thought that he was perceiving patterns in the sound of human cries. The screams of patients under the knife were not all identical, but varied according to the type of operation being performed. Amid the din of the death wails, a detached analytical mind was beginning to take shape.

In January 1839 – a mere four months after his studies had begun – he was let loose on his first dental patient: 'A boy came in looking very deplorable, walked up to me and opened his mouth. I looked awfully wise and the boy sat down in perfect confidence.' But an adolescent zeal could do nothing to conceal his inexperience. Galton began by inserting his tooth wrench the wrong way round. Once he'd found the right grip and the right tooth he took three deep breaths and pulled. The boy let out a 'confused sort of murmur something like that of a bee in a foxglove' and kicked out as Galton struggled to wrestle the tooth from his mouth. When the two of them eventually separated, Galton was dismayed to find that the tooth had broken in two. One half was stuck in his wrench, the other half was still inside the boy's jaw. Bracing himself, he went back in to finish the job, but the boy, roaring like a lion, broke free from his chair and sprinted from the building.

Galton seemed to be enjoying himself – at least he said so in frequent letters to his father – but the workload was beginning to affect his health. While the practical demands of the hospital took up most of his day, he still had to find time for his medical reading. And on top of it all he was trying to cram in evening classes in German and mathematics. Sleep deprivation, digestive problems, and stubborn headaches began to infiltrate his daily routine. His sisters were concerned, but Galton repeatedly tried to make light of his ailments. 'As to my general health', he wrote to Adèle, 'my headaches are better than they were once – a great deal better, and I have of course a little hospital fever &c., but that is all. About my mind which Lucy attacks I shall not say much, except that it is werry [sic] uncomfortable, but I shall soon get over all hospital horrors, etc., etc.'

He was so busy at work that he couldn't make it home for Christmas. By Easter there was still no let-up in his schedule, as he outlined in a frantic letter to the family: 'Can't come – quite impossible. Patients increased – awful number. Cut a brace of fingers off yesterday and one the day before. – Happy to operate on any one at home – I am flourishing – wish I could say same of my Patients.'

He went a whole year at the Birmingham Hospital without a vacation. Finally, in September 1839, he allowed himself a few weeks off in Scotland with the family. When the holiday was over he returned not to Birmingham, but to London, and King's College Medical School. Birmingham had provided some important hands-on experience for his first year of study. But now, his elders believed, it was time for a more rigorous scientific education to back it up.

London provided a much more sophisticated environment than the one he'd grown used to in Birmingham. He boarded at a grand house near St James's Park right in the centre of town, and his landlord was Richard Partridge, an eminent professor of Anatomy at King's. During the day Galton would listen to some of the country's finest scientific minds lecturing on physiology, chemistry, and anatomy, or continue with a dissection in the college basement. In the evening, he might stay at home to study the skeleton in his drawing-room closet, or talk shop with Partridge and his distinguished guests around the dinner table. Alternatively, he might forget about medicine altogether. London's attractions were not just academic. Nights out at the opera, visits to an exclusive fencing club, and the occasional evening ball all provided pleasurable distractions from his medical work.

With his social horizons expanding, Galton was in danger of turning into the perfect London dandy. But he was far too obsessed with dress codes and social conventions to play the role with any kind of conviction, too self-conscious to stand out from the crowd. Nevertheless, a busy social calendar suggested that he'd managed to replace the pressure-cooker atmosphere of the Birmingham Hospital with a more balanced and healthy lifestyle. This picture, however, may have been illusory. By December the boils were back: 'O Bessy, Bessy . . . I have had another boil exactly by the side of the former which has partially reappeared. The new one is mountainous, but alas! *not* snow-capped like Ben Nevis, but more like Ben Lomond covered with scarlet heather.'

Galton maintained a running commentary on his life through regular correspondence with his sisters. Adèle, once his favourite sister, had now fallen out of favour. He had become suspicious of her alleged spinal condition, his medical experience telling him that it was largely an attention-seeking device. With Adèle sidelined, Emma and Bessy took over the lion's share of the correspondence. Bessy was the mother figure on whom he relied for all his social and sartorial advice. Emma was the creative heart of the family, through whom he indulged his artistic fantasies. In his first term at King's College he asked Emma to paint him a couple of watercolours for his bedroom wall. His requirements were explicit: 'I should like something in a Prout style, *not* three Turks smoking their pipes in a triangle, with one blue hill in the distance and a white river between . . . but some building or other well touched up with Indian ink and reed pen.' Emma duly obliged and when he wrote again, Galton tried to quantify the pictures' importance to his life:

> I am, on average, 5 minutes dawdling in getting up, 10 minutes ditto in going to bed. During this time I must necessarily look at one side of the room or the other, and as the room is bare on the other side that can be no attraction in looking there. Therefore at least ⅔ or 10 minutes will be daily spent in looking at your pictures, besides this I am always awake about ¼ of an hour before getting up, when I cannot help seeing them as they are just before my nose, that makes 25 minutes a day, or twelve hours nearly a month or 156 hours or 19½ days of 8 hours each yearly!

In April 1840 he was busy revising for his Spring examinations, but on the thirteenth he took some time off to go and watch the Oxford and Cambridge boat race on the Thames. A paddle steamer took him up river to Putney, where he watched the race pass off without incident. On the way back, however, things went badly wrong. Returning fast with a strong tide, the paddle steamer lost control on its approach to Battersea Bridge and hit the central pier. Galton, standing on deck just behind the paddle box, was catapulted into the rotating paddle wheels and disappeared below the surface.

When he came to his senses he found himself submerged beneath the remains of the paddle box, his clothes caught on invisible hooks and nails. His hands raced over the surface of his body, seeking out the barbs that ensnared him. He finally managed to unhook himself, ease the box to one side and kick for the surface.

Back above the water line he clung to a piece of wood and floated down the Thames with the tide. He was overjoyed to see a flotilla of rescue boats navigating their way towards him. But when the first boat drew alongside, the crew seemed strangely non-committal. Eyeing up a lucrative scam, they wanted to negotiate a rescue fee before pulling him on board. Galton was in no mood to do business so he waited while the first boat moved away and made room for the next. This time the rescue came free of charge. When he was dragged out of the water, he collapsed heavily onto the deck. He was cold and exhausted. But more than anything he was deeply embarrassed. So he steadied himself, grabbed an oar, and helped row back to the river bank.

A few days later he recounted events in a letter home to the family: 'Tell Dar that if he had not taught me to swim I should have been stiff by this time and a coffin in process of being made. I am most grateful to him – and if I have children I'll make them amphibious.' The incident cost him some cuts and bruises, but otherwise he seemed unharmed. If he was suffering any kind of stress from his near-drowning experience he certainly wasn't showing it. Within days of the event he was back at college sitting his examinations, and scoring the second-highest mark out of a class of eighty students.

3

Growing Pains

In no walk of civilised life do the intellects of men seem equal to what is required of them.

Francis Galton

King's College turned out to be just another phase in Galton's peripatetic education. By the summer of 1840 he had hatched a new plan to delay his medical studies for a while and move to Cambridge University to read mathematics. Life as a perpetual student certainly held its attractions, so long as his father continued to pay the bills. But there were other reasons why the move made sense. A medical qualification required four years of study, and students had to be at least twenty-one years of age to sit for their final examinations. Galton already had two years under his belt, but if he continued with medicine for another two years he would still only be twenty. In terms of his medical degree, there were some spare years to fill.

Samuel Tertius was worried that if his son left medicine he would never return. But when Galton's scheme received the support of a well-known relative, his father may have had a change of heart. Just back from his round-the-world trip on HMS *Beagle*, Charles Darwin had been living in a house on Upper Gower Street while his half-cousin was studying at King's College. The two men had met up to chew the fat over tea and scones and Galton had come away with a wholehearted endorsement of his proposed switch to Cambridge. '[Darwin] said very truly that the faculty of observation rather than that of abstract reasoning tends to constitute a good Physician,' Galton reported to his father. 'The higher parts of Mathematics which are exceedingly interwoven with Chemical and Medical Phenomena (Electricity, Light, Heat, etc., etc.) all exist and exist only on experience and observation [therefore he said] don't stop half way. Make the most of the opportunity and read them.'

Chats with Charles were influential in other ways too. Darwin's tales of his adventures in Patagonia, the Galapagos Islands, and other exotic places must have made an impression on Galton. He had already acquired a taste for European adventure. Now he wanted more: 'In the spring of 1840 a passion for travel seized me as if I had been a migratory bird. While attending the lectures at King's College I could see the sails of the lighters moving in sunshine on the Thames, and it required all my efforts to disregard the associations of travel which they aroused.'

Once he'd completed his exams in July, he was free to travel. There was Cambridge to consider, of course, but the term didn't start until October. That gave him at least two whole months to roam. Galton had his heart set on a Scandinavian adventure to Norway and Sweden but his father, who'd be footing the bill, preferred something a little cheaper.

Samuel Tertius displayed immense generosity towards his teenage son, and would generally send him money whenever it was requested. All that he asked in return was that Galton provide weekly accounts of his expenditure. Punctiliousness in matters of pounds and pence was what you might expect from a man who had spent his entire working life in a bank. But an instinct for organisation ran throughout the family. His father, Samuel Galton, had been a stickler for figures, tables and charts, while one of his sisters had taken organisational flair into the realm of the absurd. She had systematically classified and labelled all of her pots, pans and garden tools; she also had a flower bed of twelve square feet that contained no fewer than one hundred specially painted tags.

The subject of accounts dominated correspondence between father and son, with Galton often struggling to keep up with his father's exacting demands. A letter from Galton dated 16 June 1840 is typical:

> I own that I have not kept my accounts . . . at all carefully. I have generally set my expenses down, but on scraps of paper and consequently lost them afterwards from carelessness. I do not think I have wasted any money, though I doubt if I could account for all. I am sure that I could not accurately. I don't owe anything except 32 shillings for a pair of boots and I cannot get the bill. My present riches are £14. 8s. I shall have to get a frock coat and waistcoat. The frock coat being the 3rd that I have had in London.
>
> As my journey to Norway and Sweden can scarcely be less than £50, I shall not grumble at giving it up 'in toto', but am quite ready to do so. I expect a good *row* from you by return of post, and as I deserve it, am resigned . . . It is no use on my part to blarney about 'full of contrition' and so forth, but beginning from to-day, I will send you by every Monday's post my accounts for the week preceding; and in case of omission, I wish that you would write and blow me up. Please tell me by return of post how much I am in arrears as not understanding your figures I cannot calculate it.

While Samuel Tertius expected to be taken seriously, the letter reveals how relaxed and good-natured the relationship between father and son had become. Galton often addressed his father as 'My Dear Governor', and much of his correspondence took on a gently mocking, almost sarcastic tone. 'Everything gets on capitally,' he wrote a month later, 'especially accounts. When I want to know if I have any coppers in my pocket to give to a begging crossing sweeper I do not condescend to feel but pull out my pocket-book add up and the result is sure to be correct.'

Of course, Samuel Tertius was ultimately in charge and Galton didn't always get everything that he wanted. His planned vacation to Scandinavia, for instance, was deemed too expensive. He was evidently irritated by his father's decision but tried to pretend otherwise. In the end both men agreed on a cheaper alternative: a working holiday in Giessen, Germany, to study with the re-nowned organic chemist Justin Liebig.

Within days of his arrival in Giessen Galton was writing hurried letters to his father, complaining that Liebig's methods of teaching were not appropriate to his needs. Acting on his own advice, he decided to abandon chemistry and use his time in Giessen to polish up on his German. Three days later, however, he wrote again to his father. There had been another change of plan. Forget about learning German. Forget Germany for that matter; his head was full of Byron and he was heading towards the East:

> My Dear Father,
> Being thoroughly ennuied at Giessen . . . I have determined to make a bolt down the Danube and to see Constantinople and Athens . . . I do not wait for an answer before I start for two reasons, 1st that I have not time and 2ndly as you promised me a good summer's tour to Sweden and Norway, of course you can have no objection to a comparatively civilised trip . . . So I will fancy that I have received a favourable answer, and so thank you very much indeed for your consent. My conscience being thus pacified, I will tell you something of Giessen. – It is a scrubby, abominably paved little town – cram full of students, noisy, smoky and dirty.

Without waiting for a reply Galton set off for Linz in Austria, where he hoped to catch a steamer going down the Danube to Vienna. He was in jubilant mood, caught up in the joy of his new-found wanderlust. He seemed to have shaken off some of his shyness and now other forms of passion were seeping to the surface. On an overnight stagecoach to Nuremberg he met the beautiful Marie. Unfortunately, his company was made a crowd by a Hungarian man, also intent on winning her affections. When Galton awoke to find the man holding Marie's hands, singing love songs, he moved swiftly into action: 'I then began my flirtation with much more success than my rival, at which his mustachios desponded and looked sad.'

At Linz it was obvious that the steamer that was supposed to take him down river wasn't going anywhere. Damaged during one of the Danube's capricious moods, the boat was laid up for repairs and all trips were temporarily suspended. Galton needed to find an alternative mode of transport, and fast. He had to get to Vienna to pick up the fortnightly steamer for Budapest. The details of his European journey had been mapped out with absolute precision. Minutes lost in Linz threatened to scupper his whole itinerary.

A fellow guest at his hotel offered a lifeline. Major Parry, an elderly British officer, was also in a hurry to get to Vienna. He had only the most tenuous grasp of German, but somehow he'd arranged for a private boat to take him down the Danube. He told Galton that the journey would be cheap, safe, and trouble-free, and he urged him to tag along.

In the early hours of a cold and windy morning the two men stood on the quayside waiting for their ride. Despite the gloom the river's presence was palpable. For 3,000 kilometres this European super-highway carried water, men and merchandise from Bavaria to the Black Sea. Now Galton and his military friend were about to join its cargo.

Within moments of pushing off the boat was swallowed by the river's swift-flowing current. Only then did Galton begin to have his doubts about the journey. As he looked around him he realised that the vessel he was sitting in was not so much a boat as a few rough planks loosely held together with wooden spikes. Worse still, he had entrusted his life to a crew consisting of a frail old man and a young boy. It was too late to turn round. Moving fast in mid-stream, Linz was already way behind them. This was going to be no lazy waltz down the Danube. He was going to have to work hard if his feet were ever going to find dry land again.

He was handed a long branch with a board nailed to one end. It was an oar and he was told to steer. The current was strong enough to keep the boat moving in the right general direction. But in the absence of a proper rudder there was always the danger of the boat turning broadside to the flow. In the wide, deep stretches of the river, errors could easily be corrected. But when the river turned to rapids, losing control could be much more costly. For hour after hour Galton held his makeshift oar over the stern as he struggled to keep the craft on an even keel. There was no rest. Even when he was relieved from steering duty he was down in the bottom of the boat, bailing out the water that constantly seeped through the boards. It was frantic, exhausting work and he seemed to love it. Despite all the inherent dangers of the journey he simply went about his business as if it was the most natural thing in the world. Not for the first time in his life fear focused his mind in ways that boosted his chances of survival. Twenty-four hours and one hundred kilometres after setting out from Linz, the boat pulled into the quayside at Vienna. Galton was barely a man, and a novice in the arena of travel and adventure. But in drifting down the Danube he had already earned his rite of passage.

Galton revelled in his new role as the independent traveller. In a brief note to sister Emma he could barely contain himself: 'I am laughing half the day, and I am tanned as red as mahogany, perfectly independent and in the best good humour imaginable.' From Vienna, he sent similar sentiments to his father:

It has just struck me . . . that this expedition of mine is about the coolest and most impudent thing that I have done for a long time . . . I have had the pleasantest possible voyage, nice companions – very nice indeed in some cases. N.B. Linz is universally famous for the beauty of its fair sex, and so is Würzburg, and everything prosperous. I have never enjoyed myself more . . . I would have given anything to see your physiognomies, when you received my letter from Giessen. Didn't Bessy say: 'What a monkey'?

In Vienna he reverted to a more conventional form of transport, as he boarded the river steamer bound for Budapest. In a rare effort to economise he decided to travel second class but the experience proved overwhelming: 'Fellow passengers are beastly, spitting ad-infinitum and very much crowded'. It wasn't long before he was back to the sanctuary of the first-class accommodation.

After Budapest he continued his clockwise circuit of south-eastern Europe. By mid-August he was among the mosques and minarets of Istanbul. Two weeks later he was in Athens, catching a steamer for Trieste. With time to spare he headed off to the Adelsberg caves, inland from the Adriatic coast, in search of a rare salamander, *Proteus anguinus*. These unusual eel-like amphibians live in underground streams and spend their entire lives in complete darkness. They have no eyes and the colour of their skin is, to use Galton's words, 'that of the buried portion of stems of celery'. No living specimens had ever made it back to England alive, and none probably wanted to, but Galton had other ideas. He took two *Proteus* back to Britain in a glass jar. Galton donated the unfortunate creatures to King's College. One died immediately. The other was later eaten by a cat.

In September 1840 Galton returned home to Leamington, ready to face the music. But there were no recriminations from Samuel Tertius. On the contrary, 'my dear kind father took my escapade humorously. He was pleased with it rather than otherwise, for I had much to tell and had obviously gained a great deal of experience.' Galton's progress around Europe had mapped out his own personal development. He had proved to himself that he could make it as an independent traveller, and he found it a thrilling sensation.

There were other pleasures to look forward to, or so he hoped. A new degree and a new university were awaiting him in Cambridge.

With the college term about to start there was little time to prepare. In early October, he and his father travelled down to Cambridge. It was a notable day in his life, Galton recalled in his autobiography, when he first caught sight of the dreaming spires from the top of the stage coach.

Galton was assigned rooms on the ground floor of New Court at Trinity College. It was relatively luxurious accommodation with a living room looking east into the court and a separate bedroom facing onto the trees that lined the river bank. The living room was decked out with a fireplace, sofa, table, and chairs, and Galton set about settling in to his new surroundings and making the place his own.

His first letter home suggested he was easing himself comfortably into his new environment: 'Perry gave us his first lecture today; what a pleasure it is to hear a real senior wrangler speak. My organ of veneration is so very strong that I doubt when I shall dare to address him . . . I am as happy as possible and am preparing for a long and strong pull at reading.' But the reality was somewhat different. In his autobiography he admitted that Cambridge came as a real shock to the system. 'I soon became conscious of the power and thoroughness of the work about me, as of a far superior order to anything I had previously witnessed.' For the first time in his life his own remarkable abilities suddenly looked ordinary. It was a disturbing revelation.

Cambridge University was home to some of the finest minds in the land. It was a school of academic excellence that espoused a punishing regime of study. From day one Galton struggled with the sheer volume and pace of work. In addition to attending all his lectures he was reading for ten-and-a-half hours a day. He tried to remain upbeat in his letters home, but the joyous, carefree tone of the summer had all but gone: 'Tell Bessy that there is the most extraordinary possible change in my complexion, the tan having quite disappeared. Breadth of phiz on the wane.'

Galton realised that he needed something to help him stay awake during his long, late nights of study, and his solution saw the emergence of an inventive streak. The gumption-reviver machine was one of his first inventions and one of his simplest. Essentially it was nothing more than a mobile dripping tap that soaked the head and shirt of the user. The machine proved highly effective, although it did have a serious drawback. Someone – usually Galton's college servant – had to keep refilling the water reservoir every fifteen minutes.

The dripping tap not only guaranteed many soggy, sleepless nights, it also unleashed a mechanical imagination. Galton indulged in a flurry of inventive activity during his first term. An idea for a rotary steam engine was highly original for the time, but it never got beyond the design stage. There was also a new type of oil lamp, and a lock that aroused the interest of the Bramah locksmith company. Yet none of these devices was ever fully realised, and none could match the functional simplicity of the dripping tap.

While the gumption-reviver machine was a great success, its invention was symptomatic of a more general malaise in Galton's life. The dripping tap helped him extend his waking hours, but he was still losing the battle to keep up with his studies. Only three months earlier, adrift on the Danube, and in the face of great physical danger, he had demonstrated amazing mental resilience. Now, amid the comparative calm of Cambridge, his mind was looking weak and vulnerable. For Galton, the competitive climate of college life was turning out to be far more dangerous.

By the middle of November, and barely a month into his studies, Galton gave up and went to bed. The equation was simple. Mathematics had made him ill. Bloated by an extravagant diet of trigonometry, geometry, and algebra he was forced to lie down and sleep it off. Initially, he tried to reassure his father that the problem was rheumatism and had nothing to with overwork. But it was a transparent deceit. By December he was still in bed, complaining of fever and delirium. He had no choice but to abandon his studies until the following term.

Over the Christmas vacation he seemed to recover some of his vitality, and he took a trip down to brother Erasmus's farm in the

West Country. The coach journey from Birmingham to Bristol was a memorable one, but for all the wrong reasons:

> Was shut up in the coach with a frowsy fat old gentleman and a fast young gentleman whose lungs were, judging by his breath, entirely composed of full-flavoured cubas and the Cream of the Valley. The latter was not a very pleasant companion for vinous fumes ascending into his cranium displaced what reason had existed there, and showed their presence by causing him to . . . sing a very sentimental song, and at last to open the window and afford me a very convincing proof that gin and cigars act as a strong emetic.

With Cambridge pushed to the back of his mind Galton seemed more like his old self again. His brother's farm at Loxton in Somerset was an idyllic retreat, the perfect spot to recuperate from the strain of the previous term. Galton's impressions of the landscape were recorded in a letter to his father: 'Somersetshire is really the most beautiful country I have ever seen, north of the Alps . . . and of all dull pig-headed stupid bipeds the Somersetshire clown stands pre-eminent.' The sun shone, the weather was mild and Galton was on excellent form.

He returned to Trinity College at the end of January, feeling much better for the long Christmas break. But he had still not learned to appreciate – or accept – his own limitations. In February he threw himself back into work again, only to be stopped by a relapse of his old illness. Even to a casual observer, cause and effect were only too obvious. But Galton was a boy brought up to believe in his own brilliance, and his experience at Cambridge was difficult to accommodate.

In March there were signs that he was trying too hard to compensate for his deficiencies. A travelling menagerie had come to town and in an effort to impress his fellow students he took his chance and stepped inside a cage of big cats. He survived to pen a triumphant letter to sister Emma:

Yesterday I made my appearance before the eyes of wondering Cantabs, where do you think? Why right in the midst of a den containing 1 Lion, 1 lioness, 1 huge Bengal Tiger and 4 Leopards in Wombwell's menagerie. The lion snarled awfully. I was a wee frightened for the Brute crouched so. The keeper told me that I was only the fourth that had entered that den. Nothing like making oneself a 'lion' at Cambridge. My Turkish tour and medical education does wonders and my late van Amburg performance promises to crown my reputation.

For the next two terms Galton's lion heart kept him going, but he produced only a modest performance – at least by his own exacting standards – in his end-of-term examinations. Nevertheless, by the end of his first year he seemed more at ease than he had been for a while. There was evidence that he had come to terms with his situation, learning a hard lesson in the process. 'I hope to do better in each succeeding examination' he wrote to his father, 'but ill health, for I severely overstrained myself my first term, – and I feel convinced that to have read during the Xmas Vacation would have been madness, – has necessarily kept me back.'

For the long summer vacation he joined a Cambridge University reading party of fellow undergraduates and tutors for two months of walking, talking and conspicuous consumption in the Lake District. It was his first visit to Wordsworth and Coleridge country, although Galton seemed anything but overawed: 'Windermere is said to be a beautiful lake. Wordsworth asserts that it is superior to anything abroad, but I humbly conceive that he thereby shews [sic] his patriotism rather than his taste.'

Taking a country house near Keswick, Galton was as relaxed as he had been in a long while. Whatever had been going on in his mind, he made it clear in a letter to his father that there was certainly nothing wrong with his body.

I have been deluded enough lately to climb mountains to see the sunrise, it is certainly the best regime that I know to cure romance. I for my part never felt less spiritual or more corporeal than I did when I got to the bottom of them. I had a long walk in that manner the day before yesterday. Happening to look out of the window about 12 after reading, I found that it was the most beautiful night that we had yet had. So pocketing my whisky flask and putting on my pea-coat and plaid, I walked to the town and got up a party to go, slept under a table for 35 minutes, drank some whisky punch, and then walked up Blencathra . . . stayed on the top about an hour and then got back by 7 a.m., it was about 16 miles. As the morning was splendid I then got up another party for Ennerdale. Then slept 25 minutes and walked off, and we walked the whole day, up two high mountains. I got back by 8½ p.m. and after all I really was not so very tired.

For the final few weeks of his lakeland vacation he was joined by his father and his sister Emma. Both of them rode on horseback from Scarborough – a distance of about one hundred miles – where the rest of the Galton family had been on holiday. But Samuel Tertius caught a chill on the marathon ride and, once in Keswick, his condition deteriorated with frightening speed. By the end of the month he was well enough to travel back to Leamington. But the incident introduced an atmosphere of anxiety into the whole family. It was the first time that Galton had seen his father so ill.

Samuel Tertius' health had always been held hostage by debilitating asthma. But as he got older, additional ailments began to compound the wheezing and the breathlessness. When his joints started to suffer from recurrent spasms of pain it was a sure sign that gout had joined the fray. Galton's concern for his father was obvious, but he clearly believed that gentle humour was the best kind of cure. In their extensive correspondence Galton would send up his role as the medical man with touching, playful advice:

Now my prescriptions are:

1st   That the Hospital Patient do on no occasion feel his pulse.

2nd   That the H. P. do never look in the glass to see whether his eyes are red.

3rd   That the H. P. do never examine his own health with a view to self-doctoring.

4th   That the H. P. do make improvements at Claverdon, and commit prisoners at Leamington when so inclined, but that he never attend canal-meetings, nor put himself to inconvenience or anxiety.

5th   That the H.P. do henceforth enjoy an '*otium cum dignitate*' and leave hard work to younger heads for whom it is a duty.

And now my dear Father I have finished doctoring for the present, but shall go on writing doctor's letters until I hear that you obey my rules, and that you treat your own constitution with the respect it deserves for having brought you through asthma, hard work at banking and anxieties of all sorts for so long. Indeed it is a highly meritorious constitution and fairly deserves rest.

With his father still in recovery Galton returned to Trinity College in October 1841. Once again, the start of a new term saw him in joyful, optimistic mood. He had a new mathematics tutor, William Hopkins, a raconteur by all accounts, who made mathematics fun by entering into its metaphysics. 'I never enjoyed anything so much before' went the familiar phrase to his father in November.

But the demons had not gone away. They lay dormant for a while as he struggled through his second year, but in the summer of 1842 problems re-emerged with renewed vigour. Galton was in Scotland for the vacation with another reading party. The levity of the Lake District adventure had been replaced by an altogether darker mood. The jovial banter with his father had ceased and the correspondence was all the more depressing for its frankness and honesty. 'I have been able to do but little reading since I have been here', he wrote, 'and altogether am very low about myself.'

Problems came to a head a few months later with the start of his third year at college. By now his body had begun to manifest tell-tale symptoms of strain. Dizziness and heart palpitations were frequent accompaniments to his reading. Inside his mind, things were even worse: 'A mill seemed to be working inside my head; I could not banish obsessing ideas; at times I could hardly read a book, and found it painful even to look at a printed page.' Any kind of study was impossible. His only course of action was to abandon the term and retreat to the family home in Leamington for a few months of complete rest.

His breakdown effectively ended all hope of attaining an honours degree in mathematics. But he wasn't alone. Many of his contemporaries also wilted under the intense pressure and either left the university or resigned themselves, as Galton did, to the easier life of an ordinary degree.

He returned to Cambridge the following term in a much better frame of mind. With his foot officially lifted from the academic pedal there was time to exploit Cambridge University's alternative attractions. He turned into something of a social butterfly, attending university balls, playing hockey, and engaging in Union debates. He even put himself forward for the Union Presidency, but lost the vote to a substantial majority.

Disaffected by the Union's rowdiness, he helped establish an alternative debating society, which aimed for a more genteel style of discourse. He also founded the English Epigram Society, which met three times a term. His rooms in College became a social hub, a place where distinguished guests could meet to compare cravats and sip the vintage wines supplied by his father's merchant in London.

Galton was right at home among the ruling classes. His friends were Old Etonians and Rugby boys, the sons of barons, lords and peers, the children of successful and influential figures who would themselves grow up to become pillars of the establishment. They included men like Henry Maine, later Sir Henry Maine, who went on to become a distinguished judge; William Johnson Cory, a future Master of Eton; Frederick Campbell, the eldest son of Lord Campbell, the Lord Justice of Appeal; and the future playwright, English professor, and editor of *Punch*, Tom Taylor.

Francis Galton, the Cambridge student

Perhaps Galton's most intimate friend during his Cambridge days was Harry Hallam, the son of the English historian Henry Hallam and brother of Arthur Hallam, the subject of Tennyson's *In Memoriam*. Harry Hallam was, by all accounts, a bit of a Renaissance man. '[He] had a singular sweetness and attractiveness of manner,' Galton once explained, 'with a love of harmless banter and paradox, and was keenly sympathetic with all his many friends.' With his capacity for modesty, intelligence, and effortless oration, his charms proved irresistible. Harry was not the only attractive aspect of the Hallam family. Galton also took a fancy to his sister, Julia, who was, in turn, a good friend of Emma Galton's. All four friends took a continental tour together in the summer of 1843, during the long vacation between Galton's third and fourth years at Cambridge.

As a male-only institution the university offered little immediate opportunity for meeting members of the opposite sex. Galton had to get out and about to find a girl. Judging by the accounts he sent to his father, however, his shyness still represented a serious barrier to progress:

Miss D. . . . is without exception the most beautiful etc., etc., etc. I have ever seen. I was at a hop at her Ma's house the other night (I know most of the families in Cambridge now), I was dancing with her (the daughter not the mama) . . . Today as the sun was shining beautifully I decked myself out in resplendent summer apparel, light trousers, light waistcoat . . . to make a call upon this fair creature, but as I was fast finishing my toilette, and was 'throwing a perfume over the violet' in the way of arranging my cravat ties, the wind blew and the rain fell horribly, and the streets were one mass of mud. I was in despair, but reflecting that Leander swam the Hellespont for Hero, *I* was duty-bound to *wade* as far as the Fitzwilliam for Miss D., off I set. When, however, I arrived at their door, I wisely reflected on the splashed state of my trousers before I knocked, and then retreated crest-fallen.

Galton's self-consciousness had overruled his desire and another potential encounter had ended in anti-climax. But by now he had grown used to disappointments. His whole experience at Cambridge had been one giant let-down. Arriving with such high hopes, he had become first overwhelmed, and then disillusioned with the academic treadmill. In terms of his medical career he had gained little of any practical value, but he had lost much of his health. In a letter to his father his bitterness was palpable: 'I feel more convinced every day that if there is a thing more to be repressed than another it is certainly the system of competition for the satisfaction enjoyed by the gainers is very far from counterbalancing the pain it produces among the others.' Similar sentiments echoed in the poetry that now flowed from his pen:

> Well may we loathe this world of sin, and strain
> As an imprisoned dove to flee away;
> Well may we burn to be as citizens
> Of some state, modelled after Plato's scheme,
> And overruled by Christianity,
> Where justice, love, and truth, and holiness.
> Should be the moving principle of all,
> And God acknowledged as its prop and stay.

His breakdown had exposed a reflective, spiritual side that yearned for a less materialistic society:

> How foolish and how wicked seems the world,
> With all its energies bent to amass
> Wealth, fame or knowledge.

From his embattled Cambridge perspective, life seemed a game, played for the benefit of a select few, and from which he had been excluded. There was nothing particularly unique or unusual in his frustrations, nor the way in which he chose to express them. What is interesting, however, is the extent to which he turned so comprehensively against these sentiments in later life. Eugenics, his socio-scientific philosophy of the future, would be built, according to Galton, on a solid foundation of knowledge, and exercised through a ruthless system of competitive examinations. Cambridge, you get the feeling, was far more influential than Galton would ever admit.

He drifted on into 1844, going through the motions of under-graduate life. In January he sat for his final examinations and gained his ordinary degree to very little fanfare. But he did, at least, have something to celebrate: the mathematics was now officially over. And with this heavy weight off his back, his medical career gradually came back into focus.

But circumstances at Cambridge were overshadowed by events elsewhere. Samuel Tertius had become seriously ill again. The asthma and gout that had dogged him for years now threatened a final, conclusive push. In September, and in an effort to find some respite, he went to St Leonard's on the Sussex coast. And he asked Francis, his favourite son, to accompany him.

The two men took walks along the sea front and talked at length about the future. Samuel Tertius was in good spirits, happy to be alone with Francis. 'The sea air has done wonders with me and tells every day,' he wrote to his family, 'so do not be surprised if you see my name in the papers as having gained a prize at a cricket match. Francis and myself have an occasional game at chess, but have not yet put the pack of cards into requisition.' But this was to be his last reprieve. As his health worsened he was joined by his wife Violetta, and daughters Emma and Bessy. He died in St Leonard's on 23 October 1844, aged sixty-one.

The death devastated the entire family. But for the prodigal son the loss held added significance. With his father dead, the guiding principle of Galton's life was gone, and it left him confronting a terrible dilemma. It had been his father's dying wish that his son continue with his medical career. But with his father's inheritance there was no need to work at all. If he chose to, Galton could live comfortably at leisure for the rest of his life.

4

Wilderness Years

Youths are murderous by instinct, and so was I.

Francis Galton

The boat drifted lazily on its heavy anchor chain, tracing shallow arcs against the slow but methodical movement of the Nile. Somewhere out to the east the sun was crawling out from a bed of reeds, sending splashes of orange light skipping across the water. On a sand spit up ahead, a mob of crocodiles raised their fish-filled bellies off the ground and swaggered towards the water's edge, sending a small flock of pelicans up into the air. Over on the western shore, two flamingos were in early-morning dispute. A hippo broke the surface, adding a deep, reverberating yawn to the avian duet, while a host of smaller waders loosened their stilted stance and tiptoed nervously away.

Slowly, the river stirred.

Galton rolled out of his bunk and landed heavily on the floor. He wasn't really aware of the fall. Something else was demanding his attention, a feeling that started in his toes and travelled upwards through his legs and spine all the way to his head. Perhaps the pain could have found an escape through his eyes, but his lids remained closed to the new dawn light, sealed shut by dense layers of congealed tears. His tongue tasted dry and bitter, the weathered aftermath of a decadent Egyptian night.

Splinters pierced the bare skin of his hands and knees as he shuffled out of the cabin towards the main deck. Members of the crew were already up and about, making the boat ready for the day's sail. He could make out a blur of sights and sounds, but normal sensory activity was temporarily suspended as he fumbled doggedly for the edge of the boat. With his hands gripping the gunnels, he raised his right leg over the threshold, leaned forward, and collapsed into the river.

The water hit him like a drug, as if his whole being was suddenly embalmed in the moment. For a few blissful seconds he hung motionless below the surface. But fish bigger than grown men swam in these waters, preyed upon by crocodiles twice their size. It wasn't wise to hang around too long. Within minutes he had re-emerged at the stern, using the rudder as a ladder to clamber back on board. With the worst of the evening's excesses washed away, he ordered a cup of coffee and sat down to relax.

Puffing hard on his pipe, he let his eyes roll lazily across the panorama. A movement in the water caught his attention. A large boulder seemed to break the surface about fifty metres off the starboard bow. Galton leapt off his chair and called for a gun, the biggest and meanest one available, the kind that can blast a hole the size of a football in a hippo's hide. But the hippo had other ideas. By the time Galton had the gun loaded and ready to fire the animal had sunk back below the surface. Galton blasted away at nothing in particular, a ripple here and a bubble there, while an army of spent cartridges gathered around his feet. He paused for a moment to scour the western shoreline. In the distance, maybe 200 metres away, he saw the hippo pull itself out of the water and vanish into the reeds. Galton let off one final farewell shot. The bullet missed the hippo, but it blew the head off a bickering flamingo, sending its beak cartwheeling into the sky.

In the end, Galton's decision to abandon his medical studies had been an easy one. His father's death had released him from the claw of parental expectations, and 'finding I had competent fortune and hating the idea of practising medicine . . . I determined to give it up'. Bruised by his experience at Cambridge, Galton craved freedom and a healthier, more independent lifestyle: 'I eagerly desired a complete change; besides, I had many "wild oats" yet to sow.'

The death of Samuel Tertius had precipitated the dispersal of Galton's family. Within a year, most had gone their separate ways. His mother had relocated to the country retreat in Claverdon. Daughter Emma joined her for a time, but she was an independent type who, like her youngest brother, loved to travel, and she spent months at a time away from home on European explorations. Emma had become the most sympathetic sister to Galton's character and interests, and she would remain his closest correspondent for the rest of his life. Of the remaining sisters, Bessy and Adèle were both married in 1845. Lucy's health was deteriorating, a result of the rheumatic fever that had plagued her entire life and she died in 1848, at the age of thirty-nine. Of Galton's two brothers, Darwin was already married, and living in a country house near Stratford-on-Avon, while Erasmus continued to farm his estate in Somerset.

Galton was young and wealthy enough to do almost anything he wanted. His passion was for travel and adventure. So in October 1845, a year after his father's death, he headed off to Egypt. Ever since he was a child he had shown a keen interest in shooting, so a cruise down the Nile made perfect sense. It was the ideal vantage point from which to blast away at the exotic animals that made their home along the banks of this beautiful river.

He tried to persuade Harry Hallam, his close friend from Cambridge, to join him on his trip and, initially, Hallam had sounded keen. 'I have been deliberating since I received your letter on the desirability of joining you . . . The pleasure of shooting at so large a mark as a hippopotamus of respectable size is peculiarly attractive to the mind of the infant sportsman, who like myself has been vainly endeavouring to rid creation of an orthodox number of partridges during the last month.' But in the end work commitments kept him at home. So Galton set off alone, across France and Italy towards the Mediterranean.

At Malta he caught a steamer bound for Alexandria. By pure coincidence two old university friends were also on board. Montagu Boulton – the youngest brother of Galton's great school friend, Matthew – and Hedworth Barclay had been travelling around Greece together. Like Galton, they too had their eyes on a spot of shooting and agreed to join up for a concerted assault on the Egyptian fauna.

Boulton and Barclay had already recruited some help to take the edge off their endeavours. Boulton had brought along his personal courier from Britain while Barclay had employed a Greek cook. Not to be outdone, Galton hired an Egyptian interpreter and guide called Ali. In Cairo the three men chartered a lateen-rigged sailing boat, loaded up on booze and bullets, and headed off up river. A capable crew took care of the sailing and navigation chores, while the trio of faithful servants looked after each man's personal wants and needs.

For several weeks the sailing was a breeze, as Galton and his friends enjoyed long, lazy days of luxurious living. But at the town of Korosko, 1,000 kilometres south of Cairo, the river suddenly quickened its pace, blowing away the sedentary atmosphere on board. Today Korosko lies submerged beneath Lake Nasser. But in Galton's day, before the building of the Aswan Dam, it was a place where boats going up the Nile stopped to be pulled up the rapids.

While they waited for help at Korosko, Galton, Barclay, and Boulton got chatting to the owner of the boat ahead of them in the queue. Their new acquaintance turned out to be Joseph Pons d'Arnaud Bey, a French civil engineer employed by the Egyptian authorities to survey gold deposits in the Nile's remote upper reaches. D'Arnaud Bey invited the men back to his hut, and although it was small and fairly primitive, Galton was impressed by the ordered, scientific air that pervaded the place. 'There were maps, good books and scientific instruments of various kinds, so my heart warmed towards him.'

The inspirational Joseph Pons d'Arnaud Bey

Galton was soon absorbed in the Frenchman's romantic tales of scientific exploration. Years later he claimed that this chance meeting with d'Arnaud Bey had 'important after-results to me by suggesting scientific objects to my future wanderings'. So far, Galton's travels had been in the tourist tradition, made primarily for the purposes of pleasure and relaxation. But d'Arnaud Bey talked of mapping uncharted lands, of surveying lakes and rivers, and discovering new cultures and civilisations. For Galton, it was an entirely new perspective on the possibilities of travel.

Acting on d'Arnaud Bey's advice, Galton, Barclay, and Boulton abandoned the well-trodden trail they were taking down the Nile. Instead, they hired some camels and trekked across the Nubian

Desert, heading south towards Khartoum. The landscape was a barren monotony of sand and stone, punctuated here and there by the desiccated bones of Ethiopian slaves and the husks of abandoned camels. The temperature toyed with the senses, lurching from the intense heat of the afternoons to the bitter cold of the nights. And on top of it all there was the issue of buttocks, kneaded like dough by the camel's bumping hump. You wouldn't call it a holiday but Galton wasn't complaining. He seemed to thrive on the discomfort.

Eight days later they rode into Abu Hamed, saddle sore and sunburnt and relieved to be back among the living on the banks of the Nile. But the midges soon cut short their enjoyment so they moved on to Berber, a hundred kilometres to the south. There, Galton was given a monkey as a gift by the local Governor. He bought another one to make a pair and the two primates became his constant companions for the rest of his journey.

At Berber Galton and his Cambridge chums ran into some local difficulties when they tried to hire a boat to take them further up river to Khartoum: 'The people at Berber were unruly, and so obstructive that the boatmen feared to enter with us into their own boat.' It was only when the Englishmen took matters into their own hands and set sail themselves that 'first one and then another of the men ran alongside and plunged into the water and swam to the boat and turned its head up stream'.

In the 1840s Khartoum was a sordid, sleazy kind of place populated by slave dealers and ivory traders. It was a frontier town, wild, lawless and unfriendly, and an unusual spot, perhaps, to find another old undergraduate friend. But Mansfield Parkyns was a Trinity College exile who had somehow made himself a home among the outlaws. Bizarrely, Parkyns had acquired the role of a political go-between, trying to smooth over differences between the chief tax collector of the region and the Egyptian leader, Muhammad Ali Pasha. Since the tax collector had orchestrated the murder of one of the Egyptian leader's sons his chances of success seemed slim. But the situation was complex. Abbas Pasha, the son in question, was every parent's worst nightmare, a cruel and brutal soul who seemed to have adopted Genghis Khan as his role model. Having raped and pillaged his way through the region, he finally met his match with the tax collector. Abbas Pasha wanted the tax collector's daughter for one of his wives. Although the tax collector had little choice but to agree, he exacted his revenge by paying an arsonist to torch the tyrant's tent. Wisely, the tax collector fled to Ethiopia shortly after the event, where he bumped into the itinerant Parkyns.

Parkyns cut an unusual figure. He dressed in a minimalist style, with little more than a leopard skin thrown over his shoulder. His head was completely shaven but for a Moslem tuft and his whole body was smeared in butter. Galton seemed quite taken, declaring him to have 'the most magnificent physique of a man I have ever seen'. So it was good news when Parkyns agreed to team up with his old college friends for a short trip further up river. The Nile splits in two at Khartoum, the White Nile snaking into the southern swamps of Sudan, while the faster-flowing Blue Nile travels south-east into the mountains of Ethiopia. For a sailing boat going against the current, the White Nile represented the easier path, and for the next few days they steered a course through its stagnant waters.

Out on the river again, the touring party fell back into its old and familiar groove, a life of swimming, shooting, and sunbathing, with servants always on hand to refill empty glasses. On the shooting front, however, there were signs that the alcohol was affecting their aim. For the grandson of a gun maker Galton was proving to be a pretty woeful shot when it came to hitting hippos. On one especially noisy day he blazed away at over forty of them without inflicting so much as a surface wound.

Boulton and Parkyns were especially desperate for a kill, and left the boat one night to try and track down a hippo inland. They emerged the next morning, agitated and eager to weigh anchor. While they had been waiting in ambush they had seen the outline of a large animal go down to the water's edge for a drink. Thinking they had a hippo in their sights they let off their guns. Only on closer inspection did they discover that they had just slaughtered a domestic cow. Fearing repercussions from a farmer, the boat was hurriedly turned round and sailed back to Khartoum.

The rest of the trip proved uneventful. Parkyns rejoined the lowlifes of Khartoum, while Galton, Barclay, and Boulton continued via boat and camel to Wadi Halfa, where they picked up the vessel they had left behind at Korosko. After a week in Cairo recovering from their holiday, they returned to Alexandria and said their goodbyes. Barclay went back to England and Boulton journeyed east towards Pakistan. Galton, meanwhile, was heading for the Holy Land. He, his faithful servant, Ali, and his two primate pets boarded a steamer at Alexandria bound for Beirut.

In the Near East Galton continued his peculiar, hybrid existence of rugged and rarefied living. He would disappear into the country on camping expeditions, and then return to the city for short stays in colonial villas, to wine and dine with local dignitaries and distinguished ex-pats. Basing himself in Beirut, he made enjoyable excursions to Tripoli, Aden, Jaffa, and Jerusalem. 'I lived rather stylishly,' he later recalled, 'bought 2 good horses and a pony and jobbed a native groom, Ali remaining as my personal servant'. In the long, hot afternoons, he would find some shade in one of the many public cafés, with their carefully tended gardens traversed by small streams of clear spring water. It was in these idyllic surroundings that he taught himself Arabic, quickly becoming a fluent speaker.

But in Damascus events started to turn against him. His servant Ali caught a severe case of dysentery and within days he was dead. Galton was very upset: 'I was sincerely attached to him and condoned willingly heaps of small faults in regard to his great merits.' Ali had a been a faithful and extremely honourable servant. On the long cold nights during their camel trek across the Nubian Desert, he had forsaken his own blanket to warm his shivering employer. His death, Galton acknowledged, was 'a great and serious loss'.

Galton himself was not in the best of health, plagued by a recurrent fever. In his autobiography, he claimed that he contracted the illness after sleeping near stagnant water on a camping trip in the Lebanon. But there is evidence, albeit sketchy, of an alternative and more intriguing explanation. Very few of Galton's letters from this period still exist. But reading between the lines of one extant letter from Montagu Boulton, it looks as though the origins of Galton's fever may have been a little more racy than a stagnant stream. 'What an unfortunate fellow you are to get laid up in such a serious manner for, as you say, a few moments' amusement,' Boulton wrote in September 1846. Alone and far from home, it seems that Galton may have overcome his shyness and procured the services of a prostitute. His courage may have been rewarded with a bad dose of venereal disease that would plague him, intermittently, for years to come. Wherever the truth lies, Galton's attitude towards women cooled noticeably after 1846 – the year in question – and a side of him that was already severely repressed retreated even further beneath the protective layers of his Victorian formality.

While Galton struggled with his health, Ali's death continued to hang over him. He had organised a proper Moslem funeral and sent a letter of condolence, a small present, and a week's wages to his wife in Cairo. With that, he considered the matter closed. But a few weeks later he received an official-looking letter informing him that a number of Ali's relatives were now demanding financial compensation and threatening legal action if he didn't pay up. Harassed and feeling feverish, he decided it was time to head home.

He was not alone on his long and arduous journey back to Britain. The two monkeys that had become such faithful companions on his travels throughout Africa and the Near East were still by his side. When he returned to London on a cold November night in 1846 he was unable to find accommodation for all three of them together, so he gave the monkeys to a friend to look after with specific instructions with regards to their needs. His friend passed on the information to his landlady, but she ignored it, and locked the monkeys away in a cold scullery for the night. They were found the next morning, dead in each other's arms.

Galton was happy to be home, but he had some major worries on his mind. Egypt had exposed his weakness with a weapon. The hippopotamus is one of the biggest animals in Africa. Few species present a larger target and Galton's inability to hit one was obviously a serious cause for concern.

By his own admission he was 'ignorant of the very ABC of the life of an English country gentleman'. At twenty-five, his skills at field sports were far below those expected from a man of his age and calibre. To remedy the situation Galton resolved to spend the next three years of his life honing his hunting and shooting instincts.

The pursuit began in Leamington, at the local hunt club, where men of the appropriate class and background could meet to gamble, drink, and discuss the size of their guns. The atmosphere was heady and reckless and full of killing talk. Human beings have always hunted, first for food and then for sport. But among the Leamington set the sporting philosophy seemed to have gone too far. Shooting success was measured by quantity not quality, and there were boasts of mowing down one hundred game birds in a day and eighty-seven hares in a quarter of an hour.

It would be a while before Galton could join the ranks of these gun-toting greats. He still had much to learn about the nuts and bolts of shooting. And where better to learn than a private grouse moor in Scotland owned by one of his hunt-club comrades? Throughout the late summer and autumn of 1847 he and his new Irish setter lived primitively in a makeshift shelter on Culrain moor. He was used to roughing it by now and the hardship barely registered. Besides, leisurely days among the heather did much to improve his aim.

He returned to Scotland the following year, this time to the Shetland Islands, for a solo summer sojourn shooting seals and seabirds. Not all the birds, however, were cannon fodder. Galton had hatched an ambitious plan to catch some live seabirds, for a

The English country gentleman: Francis Galton,
painted during his wilderness years

lakeland collection at his brother Darwin's estate near Stratford-on-Avon in England. After receiving some basic tuition in rock climbing he was scrambling up vertiginous cliffs to raid the nests of gulls, terns, and kittiwakes. Galton's fearlessness in the face of physical danger made him an expert climber, and when he finally left Shetland in the autumn of 1848, he was able to take with him a large crate containing a representation of the islands' bird life.

Unfortunately the crate made its long journey south on top of an unprotected railway truck, and three-quarters of the birds died of hypothermia before they ever saw Stratford. The remainder lingered on for a while but none of the seabirds settled in to their freshwater foster home. An oystercatcher was the last to die. Caught out by a particularly sharp frost one night, the bird got stuck in the ice. The next morning a fox's footprints marked the way to a lone pair of little yellow legs, rigid in the snow.

Undeterred by the failure of his seabird re-homing programme, Galton took his half-baked ideas elsewhere. He fancied a sailing adventure to Iceland, so he travelled to Ryde on the Isle of Wight and hired a cutter for a month. But he hadn't done his homework, and when he gave the order to set sail he discovered that the captain had never taken the boat beyond the Solent, so Iceland was out of the question. He didn't seem too bothered by the blunder: 'Being resourceful, I accordingly went to Lymington, and used the yacht as an hotel, getting a couple of days' hunting in the New Forest.'

For years Galton drifted from one farce to another. You could call his existence aimless but at least he was learning the ABC of an English country gentleman. The late 1840s was a period of his life marked by a mild and slightly sterile form of hedonism, the kind favoured by a rich Victorian celibate. Denying himself the pleasures of the flesh, he indulged his senses elsewhere, on riding tours in South Wales with his friend Harry Hallam, at meets with the Queen's Stag Hounds, and on shooting expeditions to Scotland. He even had a go in a hot-air balloon, but the flight ended in typically shambolic fashion when the balloon crash-landed on the front lawn of a squire's country manor. Galton was physically unharmed but he almost died of embarrassment.

At this stage in his life, still only in his mid-twenties, things could have gone either way for him. Had he followed in the footsteps of his peers at the Leamington Hunt Club, he might have frittered away his fortune on betting and gambling and then drunk himself to death. But he seemed too conscientious for that kind of fate, too keen to find some greater purpose for his life.

Clearly he enjoyed himself during these idle years, but there remained a lingering sense of dissatisfaction. His upbringing and his heritage weighed heavily on his shoulders. Deep down, he yearned for a niche in which he could define and distinguish himself. But what was his true vocation and who could help him find it? His father's death had left him free to make his own decisions. But that freedom had become an onerous burden of responsibility. Without the need to earn a living there were too many possibilities, too many choices over the direction his life could take. His existence seemed both glorious and futile, and with the passing of each idle day the load grew heavier. He needed a purpose to lead him out of this torpor, and he needed it fast. Desperate for ideas, he went to the London Phrenological Institute for a consultation with a man called Donovan.

The 'science' of phrenology was hugely popular in the nineteenth century. Essentially a more cerebral version of palm reading, its philosophy was based on the idea that a person's personality and mental faculties could be deciphered through the shape of their head, and the lumps and bumps that covered its surface. Though Galton was sceptical of the more detailed claims of phrenologists, he was always a great believer in the idea that intelligent people had bigger heads. His own head was very sizeable. Years earlier, when he was a pupil at King Edward's school in Birmingham, a visiting phrenologist concluded that Galton had one of the largest 'organs of causality' he had ever seen. Donovan confirmed the size of this organ but also supplied Galton's head with a host of other tantalising possibilities. One assessment, in particular, stood out. 'There is much enduring power in such a mind as this', Donovan wrote in his report '– such that qualifies a man for "roughing it" in colonising.'

It's impossible to say how this visit to the head doctor affected the path of Galton's future years. Donovan's words can't have come as a blinding revelation to him. He enjoyed travelling and he wasn't afraid of roughing it; he didn't need someone else to tell him that. But the phrenologist's assessment may have nudged him in a direction he was already leaning. Egypt, Sudan, and the meeting with the influential d'Arnaud Bey had turned a youthful and shapeless enthusiasm for travel into something much more focused. Now Donovan confirmed something that Galton, perhaps, hadn't the confidence to confirm for himself. Travel was his true vocation – for the time being at least.

5

The Great Trek

There is no doubt that alternate privation and luxury is congenial to most minds.

Francis Galton

Britain in the 1840s was a country on the move. Progress had become the mantra of the new Victorian age, a one-word manifesto concentrating the minds of a populace convinced of its own omnipotence. Political, social, and technological changes marked an unparalleled period of modernisation. For the first time, the nation could keep abreast of all the latest developments – as and when they happened – courtesy of the electric telegraph. In 1846, the wires were overloaded when news broke of Prime Minister Robert Peel's seminal decision to repeal the Corn Laws. The move split his own Tory party but ushered in a new era of free trade. Elsewhere, steam and iron were redefining the powers of production, while their mobile counterparts – the trains – transformed the means of distribution. By 1848 Britain had more than 5,000 miles of railway track, connecting people, places, and production like never before. These were exciting, optimistic times, provided you were neither poor, nor Irish.

Beyond British shores the infectious 'can-do' attitude was manifest in a vast and expanding Empire. Australia, India, Canada, and large parts of Africa had all become engulfed in the bosom of the mother country. Yet a map of the world in the late 1840s illustrated how much work was still to be done. Vast areas of the globe remained untouched and unseen by western eyes. Many countries were simply outlines on a surface, their internal details still waiting to be filled in. Central and southern Africa, central Australia, large parts of central Asia, and the polar regions all awaited exploration. It was a time, as Galton himself once said, 'when the ideas of persons interested in geography were in a justifiable state of ferment'.

Exploration was viewed by Victorians as a noble and patriotic profession. Explorers were the astronauts of their day, heroic pioneers, spreading Christianity, trade, and knowledge where others feared to tread. David Livingstone, the Scottish missionary, whose thirty years of African adventure began in 1840, was perhaps the most celebrated explorer of them all. But Livingstone was just one of many household names from the period. On the other side of the world, Charles Sturt was opening up the interior of Australia; John Palliser was exploring the Canadian end of the Rockies; and Richard Spruce, the Yorkshire botanist, was discovering 7,000 new species of plants in South America, including cinchona, the plant used to make the anti-malarial drug quinine. These were some of the success stories. Others were not so fortunate. In 1847 the inherent dangers of exploration were made all too evident by the exploits of another famous traveller, Lincolnshire-born John Franklin. He disappeared with 138 men and two ships somewhere around the North Pole while seeking a north–west passage to India. He was never seen again.

By the late 1840s Galton was thinking of adding his own name to the exploring tradition, and had his sights set on an African adventure. Big-game hunters like Gordon Cumming had returned from southern Africa with tales of vast grassy plains inundated with zebra, antelope, and rhinoceros. It was a sportsmen's paradise and Galton craved a piece of the action. But whereas in the past, the hunt would have been satisfaction enough, now he also wanted to have 'some worthy object as a goal and to do more than amuse myself'.

Suave: Francis Galton, the African explorer

While many parts of the southern coast of Africa had been colonised by Europeans, the vast interior remained largely unexplored. A map of the region was a half-finished sketch, still waiting for the rivers, lakes, mountains, and deserts to be added. In 1849 Livingstone, Oswell, and Murray drew in a few significant details when they ventured north from the Cape, crossing the Kalahari Desert and discovering Lake Ngami in what is now Botswana. A few other brave souls had lifted the lid on remote parts of Namibia to the west. But by and large this area remained a blank canvas to outsiders. Opportunities for exploration were everywhere.

If Galton was going to distinguish himself as an explorer he needed more than good intentions. The credibility of his venture also depended on the backing of an authority that would help him realise his objectives, and lend his achievements some legitimacy. Fortunately Galton had both friends and relatives with connections to the appropriate circles. Chief among these was his cousin, Captain Douglas Galton, a Royal Engineer and a fellow of the influential Royal Geographical Society.

Douglas Galton ensured that his cousin gained introductions to all the right people, including the colonial secretary of the Society, Lord Grey. Discussions with the Society helped Galton turn his vague ideas for African adventure into an officially endorsed itinerary. And while the Society could not furnish any financial support, it could provide letters of introduction designed to ease the burden of bureaucracy, and make his African travel run smoother.

As Galton made his preparations, Sir Hyde Parker, a friend from his shooting days in Scotland, urged him to take on a companion as a second-in-command. Parker already had someone in mind when he made the suggestion. Charles J. Andersson was a Swede who had been brought up by the English writer and Scandinavian sportsman, Charles Lloyd. Supremely fit and agile, Andersson had been 'sent to England to push his way to fortune as best he could. His capital wherewith to begin consisted of a crate of live capercailzie, two bear cubs, and the skin of one of their parents.' Galton considered his character ready-made for the expedition, and he was immediately signed up for the voyage.

In early April 1850 Galton and Andersson assembled themselves and their equipment at Plymouth, where the boat that would carry them to Africa was waiting. Galton had booked berths on the *Dalhousie*, a teak, three-masted East Indiaman. The ship provided cheap-rate travel for emigrants to the Cape and accommodation for most of those on board was primitive. But there was also a small cabin-class section, where conditions were less cramped, and which offered exclusive use of a high poop deck. This would be Galton's home for the next three months.

On 5 April the boat was ready to set sail, and Galton had just enough time to rush off a last-minute missive to his mother: 'At length we are off . . . The weather has been very wild here, but has now reformed. Good bye for $4\frac{1}{2}$ months when you will get my next letter.' But Galton's prediction proved premature. A month into his journey the *Dalhousie* passed a ship on its way back to Britain and Galton was able to send an update a little earlier than expected. Excerpts from the letter provide a glimpse of life on board:

> At starting we had very bad weather and were about 10 days in the Bay of Biscay. A poor girl, a passenger, a clergyman's daughter, who was going out to settle with her brother at the Cape caught a severe inflammation of the lungs and died there. Our passengers make a very amusing party, and the time passes as pleasantly as can be. I am quite a good hand at taking observations and have learned about 600 Sichuana words (the language I shall have to speak). Andersson is a very good fellow. I keep him in excellent order. He rammed a harpoon almost through his hand the other day, but he is a sort of fellow that couldn't come to harm. He had an old gun burst whilst firing it last week and only shook himself and all was right.

The long sea voyage gave plenty of opportunity to prepare for the adventure ahead. While Andersson perfected his muscle tone with extravagant gymnastic routines high up in the rigging, Galton honed his navigational skills, familiarising himself with the map-making principles he would use in charting the unknown lands of the African interior. He had left England with no formal instruction in the use of sextants and compasses, so everything had to be self-taught on board ship. But he took 'very kindly indeed to instrumental work' and became as capable with a sextant as it was possible to be.

It took more than eighty days for the *Dalhousie* to make its way from Plymouth to Cape Town. Unable to sail directly into a head-wind, the ship had been forced to steer an erratic course which, at one point, took it within sight of mainland South America. It was a winter's day in June when the welcoming site of Table Mountain came clearly into view, and crowds of emigrants gathered on the lower decks, keen to catch the first glimpse of their new home.

While life at sea had been relatively serene and trouble-free, it was a different story on land. As soon as Galton set foot on the Cape Colony he was greeted with the news that political turmoil threatened to scupper his entire travel itinerary. His original idea had been to head north from the Cape towards Lake Ngami, in Botswana, and then explore the uncharted areas to the north of the lake. But now, he learned, a barrier of Boers was standing in the way.

Unsettled by British colonial interests and reforms, the Boers, farmers, and descendants of the first Dutch colonists on the Cape, had uprooted themselves and headed out on their Great Trek north in 1835. By 1850 relations between the Boers and the British government had deteriorated to such an extent that Galton could not be guaranteed safe passage north of the Orange River.

Undeterred, he hastily rearranged his travel plans. Instead of exploring the area to the north of Lake Ngami, he would look to its west, to the vast tract of uncharted land that lay between the tropic of Capricorn in the south and the Cunene River in Portuguese-controlled Angola to the north. He could access this district – and circumnavigate the Boers – by taking a boat from Cape Town to Walfisch Bay, a remote trading post some 400 miles north of the

Orange River, on the edge of the Namib desert. A few isolated mission stations had already been established inland of Walfisch Bay, providing useful stopping-off points for his journey towards the interior.

The proposed route was not without its dangers. Apart from the obvious difficulties of finding enough food and water, it was the locals who presented the most serious concern. Nama people in the south of the country were effectively at war with the Damara tribes further to the north, and an air of anarchy and distrust pervaded the whole region. Cape Town was full of grim tales of Nama atrocities. Fortified by European guns they would make devastating raids on Damara villages, murdering and mutilating the inhabitants before making off with their oxen. The cattle were then sold on to Dutch traders in the south in exchange for more firearms.

All this was a huge embarrassment to the British government, who were keen to establish good relations with indigenous Africans and distance themselves from the more oppressive and hostile stance of the Boers. Galton's visit offered the chance of some much-needed bridge building. Sir Harry Smith, the Governor of the Cape, hoped that Galton might act as a kind of cultural missionary, going into the region and preaching the reassuring gospel of the Empire. To help him on his way he provided Galton with an official letter of explanation, written in English, Dutch and Portuguese, and stamped with a red seal so large that it needed to be carried in its own tin can.

Galton had also brought along his own instruments of appeasement, a vast collection of flotsam and jetsam with which he intended to trade and barter his way through difficult country. There were guns, beads, knives, looking glasses, accordions, hunting jackets, military uniforms, swords, bracelets, yards of picture chains, and pounds of tobacco. Pride of place went to a theatrical crown he had bought in Drury Lane, and which he hoped to place 'on the head of the greatest or most distant potentate I should meet with in Africa'.

With his travel plans settled, Galton took time out in Cape Town to learn as much as he could about the land that lay ahead, and gather together the equipment and personnel he would need for his journey. For weeks he paced the city streets, hunting down cooks, interpreters and wagon drivers; dogs, mules and cattle; food to feed them all and a boat to carry them in.

By the second week of August he was ready to set sail. 'An hour more and I shall be off,' he wrote to his brother Darwin, 'with I think nearly as efficient a lot of men and cattle, as could possibly be met with.'

I have been obliged to freight a small Schooner to take me and my traps about 1,000 miles up the Coast to Walfisch Bay where I land and go towards the interior . . . Andersson is a right good fellow, and he does whatever he is told, which is particularly convenient. My head-man is one of the best known servants in Cape Town. He is Portuguese – has travelled all his life, speaks Dutch and English perfectly and has always been liked by everybody. Then I have a Black, to look after my nine mules and horses. He calls me 'Massa' and that also, is very pleasant. He is a tall athletic well built fellow, who has worked uncommonly well in Kaffir land. Next comes a smart lad to help him, and then I have 2 Waggon drivers and 2 leaders for the Oxen . . . Our party therefore consists of seven servants, Andersson and myself . . . I have been obliged to lay in a very great quantity of stores, for the place where I am going to land, has no communication established with any other port, and nothing that is forgotten can be replaced . . . The country is utterly unvisited by any White, after the first 300 miles, no traveller or sportsman has ever been there at all, only a few missionaries and traders – but the universal account from the Natives is, that the further you go, the richer it becomes.

It took a little under two weeks for Galton to make the 1,000-mile trip north to Walfisch Bay. On the way, he was joined by a convoy of humpback whales. 'It was a magnificent sight; for the whole sea around us was ploughed up by them.' The Bay was named by the Dutch after the 'whale-fish' that populated its waters during the breeding season. The sea here was rich and fertile, attracting hungry fish in their billions. Pelicans hovered overhead, while closer to shore great flocks of flamingos trawled the shallows for shrimps and other crustacean titbits.

Next to this bounty, the land was a smouldering furnace, too dry, hot, and monotonous for life to put down its anchor. It was an intimidating sight, the low sandy shore 'dancing with mirage' and painting a picture of the 'utmost desolation'. But life was to be found here. There was a small storehouse on the beach and a group of men standing nearby, their matchstick figures vibrating in the heat.

Galton hastily arranged a landing party and headed for the shore. As the boat hit the beach the local Nama tribesmen, seven in all, came down to confront them. The men were dressed in an odd assortment of trousers and animal skins. Three of them had guns, and there were some uneasy moments as the two parties eyed each other up. 'They drew up in a line,' wrote Galton, 'and looked as powerful as they could; and the men with guns professed to load them.' But a bit of low-key bartering soon broke the ice. Galton handed over cotton and tobacco in exchange for milk and meat. Water, however, was his main concern, and after some improvised sign language he was being led inland, through heavy sand and along the path of a dry river bed.

The conditions were intense. August was mirage season, a time when pelicans could be 'exaggerated to the size of ships', and the whole ground waved and seethed 'as though seen through the draft of a furnace'. The bed of the river was bordered by high, shifting dunes and cut an unruly course through dense bushes infested with ticks. Eventually, Galton was delivered to a pool of green and stagnant water about fifteen centimetres in diameter. Sand Fountain, as it was called, was the only water for miles.

The nearest permanent habitation, and Galton's immediate destination, was Scheppmansdorf, a German mission station about twenty-five miles inland. A messenger was dispatched into the dunes to go and fetch Mr Bam, the resident missionary. The next morning two men appeared riding saddleback on a pair of oxen. One was Mr Bam, the other was Mr Stewartson, a cattle trader who had fallen on hard times and now eked out a living at the station. Bam was warm and sociable but full of foreboding over the prospects of Galton's expedition.

That night Galton's two horses decided to make a run for it. Galton wasn't aware of the break-out until morning, and in the panic that followed he foolishly sent two of his men into the desert to look for them. Bam, more wise to the ways of the wilderness, knew how easy it was for the inexperienced to lose their way among the sand dunes. So by mid-morning, and with the men still missing, two locals were sent out to track them down. Later that day horses and men emerged dehydrated from the desert. This time they were lucky, but the incident served as an early warning for Galton. He would need to be fully alert to all possible dangers if his expedition was going to succeed.

Getting equipment off the ship and onto the beach was never going to be straightforward. The only access to the beach was by small dinghy, so almost everything had to be broken into its constituent parts before it could be carried ashore. The horses and mules, too big for the dinghy, were forced to brave the shark-infested waters and swim for it. In the confusion accidents were inevitable. Some things ended up in the sea. On land, a New-foundland dog that Galton had rescued from certain death in Cape Town took fright at the sound of wagon whips. It sprinted off into the desert and was never seen again.

With his cargo piled up on the beach Galton penned a quick letter to his mother, his last for many months. 'At last I am fairly on the desert with everything before me quite clear and apparently easy,' he wrote.

> I have, I find, made a most fortunate selection of men. They work most willingly and well, and nearly all know some kind of trade. . . . Ostriches are all about round here, though I have seen none yet. We got five eggs and ate them the other day. Lions infest the country about 30 miles off; if they don't eat my mules I shall have delightful shooting. The ship unexpectedly is on the point of starting, so my dear Mother, a long intended letter is spoilt . . . you will hear no more from me, for I fear a very long time.

The schooner sailed away, taking the letter with it, and leaving Galton with his men and mules and a mountain of equipment in a tidy shambles on the shore. There was no going back. Ships were rare in these parts, and it might be a year before another one was seen in Walfisch Bay. But Galton was in buoyant mood and eager to make a start. Men were put to work, and out of a muddle of axles, wheels, and wooden planks the recognisable form of the wagons began to take shape.

All the equipment would have to be ferried over the soft sand to Scheppmansdorf, a tedious undertaking that Galton delegated to Andersson. This left Galton free to pursue more important matters. He had received news from Bam that a rogue lion was causing mischief in the Scheppmansdorf area. Leaving Andersson behind at the watering hole, he rode off to the mission station, ready to do battle with the king of the beasts.

Scheppmansdorf was built on an island within the dry bed of the Kuisip River. A whitewashed church took pride of place in the centre of the island and divided the Bam and Stewartson cottages from the twenty or so huts of the locals. Three acres of pasture and gardens were watered by a small stream that ran through the village. At night a herd of oxen would wander in from the fields to settle down and sleep among the outbuildings. Galton was quite taken by the place, especially the delightful Mrs Bam: 'I am sure that Missionaries must find great favour in the eyes of the fairer sex, judging from the charming partners that they have the good fortune to obtain.' Galton's libido may have been withdrawn from active service after his unfortunate experience in Beirut. But here was ample evidence that it was not altogether moribund.

On his second night in Scheppmansdorf Galton was woken by a racket. Above the din of barking dogs and the melancholy groans of the oxen, he could hear people running about, screaming for their lives. Feeling hungry, the lion had come into the village to feed. Alone in his bunk, all Galton could do was pull his covers a little higher and try and block out the bedlam outside. 'I was glad', he confessed, 'there was a door to my outhouse, for fear the lion should walk in.' By morning it became clear that the lion had made off with one of Stewartson's dogs, and a hunting party was hastily arranged. Bam and Galton took horses, Stewartson and two of Galton's men rode oxen, while a stream of hangers-on followed on foot.

The hunting party tracked the lion for eight miles before catching sight of him. This king of the beasts had obviously had his day. He was a sad specimen, starving to death in the quiet solitude of the sand dunes. Galton, still on horseback, pulled out his rifle, lined up the lion in his sights, and fired. A messy spray of flesh and bone exploded from the lion's rear end as the bullet burrowed its way deep into the animal's back. The lion stumbled on for a while, its life draining away. Bam put it out of its misery with a single shot to the head. When they cut open the animal they found Stewartson's dog inside, in five, barely digested, pieces. As Galton so astutely observed, 'The meal must have disagreed with him.'

The mission station at Otjimbinguè lay about one hundred miles north east of Scheppmansdorf, on the banks of the Swakop River. Finding the mission station wouldn't be difficult once Galton had found the river. But finding the river was another matter alto-gether. Scheppmansdorf was separated from the Swakop by the vast Naarip Plain, a dry, flat, grassless void thirty miles across. At night dense fogs settled down on the plain, reducing visibility to zero and making navigation almost impossible. Losing your way wasn't the exception; it was the normal course of events.

On 19 September 1850 the expeditionary force left Schepp-mansdorf. Stewartson, mounted on his ox, led the way, while Galton, Andersson, and the rest of the team followed with horses, mules, and oxen. Despite their fears, the fog stayed away for their sixteen-hour trek and clear night skies illuminated a straight course to the Swakop.

The river had cut a valley that sank 1,000 feet below the plain. From the ridge the party made the steep descent to the bottom of the gorge, where a small trickle of water ran through a green and glistening oasis of tidy lawns, reed beds and camel-thorn trees. After the gruelling march across the desert, it looked like an ideal place to relax. But while the men bathed in the rock pools, and the animals made the most of the grazing, Galton was fretting about food, or the lack of it. He had pinned his hopes on finding wild game within the Swakop basin. But for all the lush vegetation there was scant evidence of any wildlife – at least of the edible kind. He had brought along half a dozen goats as an emergency back-up. But by the end of his first evening on the Swakop, they were already inside the bellies of his hungry crew.

The next day they packed up and headed east, following the ridge of the Swakop canyon. Away from the coast and the cooling sea breezes, the excessive temperatures of the interior were begin-ning to take their toll. Some of the men, confused by the powerful sun, became separated from the rest of the party. They rolled into camp later that evening, minus three of the mules. Overwhelmed by the heat, the animals had sat down and refused to budge. Fearing for their own lives, the men had no choice but to leave them behind.

That night's camp was established on a ridge overlooking the Swakop. Stewartson suggested sending the remaining mules and horses down into the canyon to feed. Trusting in his better judgement, Galton agreed to let the animals go and he went to bed in 'happy ignorance of the fate that awaited them'. The next morning, when Galton and his men went down to retrieve the animals they found, to their horror, 'not a mule or a horse, but their tracks going full gallop in a drove, and by their side, the tracks of six lions, full chase'. Following the trail they came across a lion feasting on one of Galton's chestnut mules. A little further on a horse lay dead on the ground.

Two was the sum of the slaughter. The remaining animals had scarpered down the Swakop and were eventually recovered. But this was another blunder that Galton could have done without, and the events cast serious doubt over Stewartson's credibility. The one consolation was that the lions had left behind enough prime cuts of mule and horse meat to keep the crew going for a few more days.

Galton decided to abandon his route along the ridge, preferring instead to stay down in the valley, where there was always water and pasture for the weary animals. But lions continued to haunt the caravan. Everyone had trouble sleeping during the long, dark nights as their roars echoed around the canyon.

Lions were not the only thing on Galton's mind. The scarcity of game continued to worry him. The absence of large grazing animals didn't seem to make any sense. The Swakop looked like the perfect place to find herds of buffalo, antelope and zebra. But the valley floor had been emptied of its animals. This was a wilderness without wildlife. Something or someone had driven it away.

Hans Larsen was the man to blame. An ex-sailor and cattle trader, Larsen earned a living at the Otjimbinguè mission station, rearing cattle and doing odd jobs for the missionaries. In a seven-year shooting binge this trigger-happy Dane had systematically wiped out all of the Swakop's large grazing mammals. Larsen was the kind of fellow who would have been welcomed with open arms back at the Hunt Club in Leamington, parental credentials permitting. It is no surprise, therefore, to find that Galton signed him up to the expedition as soon as he reached Otjimbinguè.

Galton's arrival in Otjimbinguè was greeted with news of an atrocity at Schmelen's Hope, an outlying mission station fifty miles further up river. The missionary and his wife had managed to escape to nearby Barmen, but many of the villagers had been massacred by a rampaging mob of Orlams. The perpetrator was well known to everyone. His name was Jonker Afrikaner.

Conflict in the region had been dragging on for decades. But what had started out as minor territorial disputes over grazing land had developed into something much more sinister. To the north of the Swakop River was Damaraland, a dry and barren place populated by the Damara people, nomadic cattle traders who had been pushing south in search of new pasture for centuries. To the south of the river lived the Nama tribes, pastoral farmers who had been displaced northwards by Dutch and British colonists in the Cape.

For decades the Damara and Nama people had been engaged in low-level conflict. But in the 1830s the balance of power suddenly shifted with the arrival of Jonker Afrikaner and the Orlam people. The Orlam were descendants of the illegitimate children of Nama slaves and their Afrikaner masters. Rebelling against the oppression of their slave owners, they had moved north into Namaqualand. Armed with European guns, the Orlam and Orlam-backed Nama had quickly subjugated the Damara tribes. Trade networks with Europeans in the south allowed the Orlam to maintain their dominance. Using commando-style tactics they would raid Damara villages, steal their cattle and send them south in exchange for weapons.

Jonker Afrikaner was chief among the Orlam. He had established a home for himself at Eikhams (now Windhoek, the Namibian capital), from where he orchestrated his military manoeuvres. Although hostilities were directed primarily towards the Damara people, Jonker took umbrage at anyone who got in the way of his interests. He understood the value of the gun and much of the violence stemmed from his struggle to control the trade in guns and cattle. Jonker's tactics had engendered an air of lawlessness throughout the region. Damara, Nama, and Orlam were all engaged in a seemingly endless cycle of tit-for-tat violence and robbery.

Galton's proposed route would take him right through the heart of Damaraland. But neither Jonker nor the Damara had any interest in allowing him access to the region. Jonker suspected that Galton might supply the Damara with guns, threatening his trade monopoly and authority, while the Damara thought that Galton was an Orlam spy.

Faced with such concrete opposition to his travel plans, Galton decided that the best way forward was to broker some kind of peace between the rival factions. After only a day in Otjimbinguè, he saddled up his ox and made the short journey east to Barmen, the next mission station on the Swakop, and an intelligence hub for Damara and Nama movements.

At Barmen Galton witnessed for himself the kind of cruelty that had been dished out to the victims of Schmelen's Hope. He met two Damara women, 'one with both legs cut off at her ankle joints, and the other with one'. The Orlam had hacked off their feet so that they could get at the bracelets they wore around their ankles. The women had stopped the bleeding by digging their stumps into the sand, and then crawled twenty miles from Schmelen's Hope to Barmen. There were other stories of appalling brutality. Jonker's own son, for instance, 'a hopeful youth, came to a child that had been dropped on the ground, and who lay screaming there, and he leisurely gouged out its eyes with a small stick'. The overall death toll was impossible to estimate. Hyenas devoured the dead before anyone had a chance to make a body count.

Jonker's behaviour was a huge embarrassment to the Cape

government. As he had been born in the colony under British rule he was effectively a British subject. The British were keen to expand from the Cape colony and establish trade links with the Namibian interior. But to do so effectively required political stability and the trust of the Damara people. If for no other reason than the interests of good business, Jonker had to be appeased. Dutifully, Galton penned an indignant letter to the Orlam chief, spiking the usual diplomatic niceties with some barely concealed threats:

> I wish strongly to urge you on the behalf of common humanity and honour to make what amends you can for your late shameless proceedings. Your past crimes may profitably be atoned for by a course of upright wise and pacific policy, but if the claims of neither humanity, civilisation or honour have any weight with you perhaps a little reflection will point out some danger to your personal security.

While he waited for a response Galton killed time by learning a little more about the Damara and their language, making small talk with the missionary and his family, and busying himself with his sextant and compass, mapping out the features of the surrounding land. In his role as surveyor, his basic task was to measure the positions of geographical landmarks he encountered on his journey. Triangulation became his standard method for mapping the prominent features of the landscape. The tops of three hills, for instance, became the three corners of a triangle, whose angles he could measure with a compass and whose sides he could deduce via trigonometry. As he travelled he would often make lengthy detours to take compass readings from distant hills or mountains so that he could incorporate as many features as possible into his map.

On clear days and nights he used his sextant to measure the angle of inclination of various celestial objects – the sun, moon, and some of the better-known stars were common targets – and, armed with the appropriate tables, convert these angles into the latitude and longitude of the spot on which he was standing. This kind of global positioning became part of Galton's daily routine and he made literally thousands of measurements along his route.

As he moved inland he added other dimensions to his map of the physical landscape, often with only the most primitive pieces of apparatus. By boiling a kettle of water, for instance, he could estimate his height above sea level. Because air pressure decreases with altitude, the boiling point drops the higher you go. So he could work out his altitude simply by measuring the temperature at which the water boiled.

Galton took on all these map-making tasks with a real enthusiasm. He seemed to delight in the accuracy and attention to detail that good surveying demanded. 'I have been working hard to make a good map of the country,' he wrote to his brother Darwin, 'and am quite pleased with my success. I can now calculate upon getting the latitude of any place, on a clear night to three hundred yards. I have fortunately got very good instruments and have made simple stands to mount them upon, so that I can in a few minutes set up quite a little observatory.'

He took great pride in his navigational instruments and rarely went anywhere without his sextant. But it was only in the missionary village of Barmen that he came to realise just how versatile the instrument really was. The celestial object this time was no moon or star in the sky, but the venerable Mrs Petrus, a woman who was, in Galton's words, a 'Venus among Hottentots'.

The sight of Mrs Petrus seemed to trigger some kind of hormonal thunderstorm inside Galton's dehydrated body. He was intoxicated by the physical form of the woman, by the voluptuous curves of her breasts and buttocks: 'I was perfectly aghast at her development, and made inquiries upon that delicate point as far as I dared among my missionary friends.' But it was with his brother, Darwin, that he shared his most intimate thoughts. The letter home, his first for many months, revealed the true extent of his passion:

> I am sure that you will be curious to learn whether the Hottentot Ladies are really endowed with that shape which European milliners so vainly attempt to imitate. They *are* so, it is a fact, Darwin. I have seen figures that would drive the females of our native land desperate – figures that could afford to scoff at Crinoline . . . Had I been a proficient in the language . . . I should have said that the earth was ransacked for iron to afford steel springs, that the seas were fished with consummate daring to obtain whalebone, that far distant lands were overrun to possess ourselves of caoutch-ouc [rubber] – that these three products were ingeniously wrought by competing artists, to the utmost perfection, that their handiwork was displayed in every street corner and advertised in every periodical but that on the other hand, that great as is European skill, yet it was nothing before the handiwork of a bounteous nature.

Being a scientific man, Galton was 'extremely anxious to obtain accurate measurements of her shape'. Getting hold of her vital statistics, however, would be difficult. He was far too bashful to approach her directly, and besides, he didn't speak a word of the local language. He was brooding over his predicament one day when he spotted Mrs Petrus standing in the shade of a tree, looking more beautiful than ever. Suddenly he had an exquisite thought. There was no need to approach her; he could flirt from a distance with his sextant.

Holding his instrument in his trembling hands Galton gradually brought Mrs Petrus into his line of sight. Slowly and carefully he began to map the audacious curves of her body. When Mrs Petrus tired of the attention and moved away, Galton whipped out his tape and measured the distance to where she had been standing. Later that night, alone in his bedroom, he 'worked out the results by trigonometry and algorithms' using simple mathematics to conjure a vivid image of Mrs Petrus.

With his urges satisfied, Galton tried to clarify his travel arrangements with the locals in Barmen. One Damara man told him of a great lake called Omanbondè, lying on the northern edge of Damaraland. Galton had his doubts about the accuracy of the information since the same man also talked of a strange group of people who were deficient in elbows and knees. Lacking articulated arms, they were unable to feed themselves, so they had to eat in pairs, each person lifting food to the mouth of their partner.

Even if the man was telling the truth, and the lake did exist, it was difficult to predict times and distances. The Damara people paid little attention to European conventions of time and didn't count beyond three. One man told Galton that if he started for Lake Omanbondè, and travelled as fast as he could, it would take him so long that he would be an old man by the time he got back.

Before he could travel anywhere Galton had to get his animals in order. Horse distemper had decimated what was left of his mules and horses and he was relying, increasingly, on oxen. The ox was the standard vehicle of the African caravan. Like horses, they could be saddled and ridden, or used as pack animals. But their natural gregariousness made them more reliable travelling comrades than horses, always keen to stick close by the camp fire at night. Galton had been highly sceptical of the ox when he first arrived in the country. But after months in their company, he had grown quite attached to them.

In a country without cash, oxen were the prized commodity. 'A loose ox in Damaraland,' Galton observed, 'is as quickly appropriated as a dropped sovereign in the streets of London.' The Damara considered the ox as an object of reverence and their ability to identify individuals among a herd was remarkable. Each time Galton's entourage arrived in a new village the locals would quickly scan his herd to make sure that all the animals were his own.

Galton had bought about fifty oxen from Hans Larsen, but most of them were not ready for work. Oxen require some basic education before they can go out on the road. Like horses, they need to be broken in to a working life. Foolishly, Galton decided to take on this task himself.

The first animal to enter the corral had a look that was the very definition of obduracy. For the first hour it skipped and danced around the ring, avoiding all efforts at entrapment. Sometimes the ox would fall still and look for grazing on a non-existent piece of grass. Sensing his chance, the man with the rope would close in from behind, only to disappear in a cloud of dirt and dust as the ox went buckaroo. When a rope was eventually secured around a rear leg the ox threw itself down on the ground and broke its thigh bone. Galton had to shoot it.

Ox number two proved more frisky than the first. It sprained its ankle and then 'got savage, and chased everybody, running upon three legs'. The ox took refuge in a patch of bushes next to a large nest of hornets. 'Between the mad charges of the animal and the stings of the hornets, we were fairly beaten off,' Galton lamented, 'and had to leave him the whole day by himself.' Despite the setbacks Galton remained philosophical. The character of the ox, he declared, 'is totally different from that of a horse, and very curious to observe; he is the infinitely more sagacious of the two, but never free from vice'.

It was a month before there was any word from Jonker Afrikaner, and his letter, when it came, was a disappointment. It was long and rambling with few concrete assertions. Conciliation seemed to be the last thing on his mind. This suspicion was confirmed when, days later, news came of fresh outbreaks of violence in the region. Refugees began streaming into Barmen from outlying areas, bringing with them more harrowing accounts of Orlam atrocities.

Angered by Jonker's indifference to his travel plans, not to mention the wishes of the Cape government, Galton decided that a change of tack was needed. Conventional diplomacy had obviously failed; it was time for direct action. He was going to ride down to Eikhams and confront Jonker face to face.

Local chiefs in the area were prepared to be pacified, but get on the wrong side of them and the consequences were usually unpleasant. Chief Umap, for example, had reacted badly to the recent death of his son from an illness. Believing that his life had been charmed away by local Bushmen, he had rounded up eight random suspects. A fire pit was dug and the suspects thrown in. The hole was then filled up with hot earth and another bonfire built on top for good measure. As Galton so diplomatically put it, Umap was 'a very respectable Hottentot; but he is classed as one of the old school'.

Galton took only three men with him on his forty-mile trip south to Eikhams, and it was dusk when the four pulled up on the outskirts of the town. As they surveyed the scene, even Galton's ox caught the excitement, 'and snuffled the air like a war-horse'. When they were ready, Galton kicked them into a canter. Galton was dressed to kill, kitted out in hunting cap, scarlet coat, cords and jack boots.

A crowd was gathering as the oxen homed in on the largest hut in town. Galton took the lead, the other three following close behind. As he got nearer he saw that Jonker's hut was protected by a four-foot moat, but his ox took it on, jumping the ditch like a steeplechaser. When the animal came to rest, its head 'not only faced, but actually filled the door of the astonished chief'. Jonker had been sitting on the floor, quietly smoking his pipe, when Galton burst in. He could only watch in silence as Galton gave him an earful, waving around the government order he'd been given in Cape Town for added effect. The grandstanding performance seemed to pay off. 'I made the man as submissive as a baby', Galton later bragged in a letter to his friend Frederick Campbell.

As Galton retired to bed for the night Jonker was busy sending out generous slices of humble pie. Over the next few days he openly confessed his sins and promised a more peaceful future. But this message had to be circulated throughout the region, and Galton spent weeks travelling around neighbouring villages in an attempt to get all the tribal chiefs to agree to a common aim. Without wishing it, he found himself in the role of chief lawmaker, drawing up a basic moral code of conduct by which all Nama people should abide.

With local difficulties ironed out and chiefs forewarned, Galton was finally ready for his journey into Damaraland. Progress had been painfully slow. It had taken seven months to cover the 220 miles from Walfisch Bay to Schmelen's Hope. While it was true that local politics had slowed him down, it was the wagons that were most at fault. For weeks they had stood static as new teams of oxen were trained to pull their heavy loads. And when their wheels did finally start to turn, they had been hampered by the soft sand and undulating terrain. But for all their inconvenience the wagons were a necessary burden in a land built on the economic principles of exchange and barter. Loaded with iron, guns, clothes, and trinkets they were, in effect, Galton's bulging wallet.

On the eve of his departure Galton was in good spirits. 'We have had admirable health', he wrote to his mother, 'and now although the sun is high yet the rainy season has brought its clouds and the climate is really very pleasant.'

I am becoming a stunning shot with my rifle, and always shoot plenty of ducks, partridges and guinea fowl with it . . . If fortune favours me, I shall be able, I have no doubt, to make an entirely open road for future commerce here – where people may travel and trade without any danger. I have taken great pains about mapping the country. – It is a great amusement, and the Government at the Cape, expressed so much anxiety about creating a cattle commerce here, that I have no doubt that what I have done will be soon followed up.

On 3 March 1851 the caravan made its way out of the ruins of Schmelen's Hope, heading north towards the real or imagined lake of Omanbondè. The touring party had swollen to twenty-eight men. In addition to Larsen, Andersson, and those he had hired in Cape Town, Galton had taken on a gaggle of local guides and general helpers. But animals outnumbered people by more than four to one. Oxen made up the bulk of the beasts, with ninety-four in total. Some pulled the wagons, others carried packs or riders, while an unfortunate few did little but carry the meat on their bones. Twenty-four sheep – also part of the mobile larder – completed the caravan. Together, this itinerant community of men, meat and machines heaved itself along the broken ground of the Swakop basin onto land never before touched by European feet.

The caravan soon left the valley behind as it made its way up and out onto a high desert plateau. Here the ground was sandy but firm. It would have made an excellent surface for the wagons had it not been for the thorn bushes that thwarted all attempts to travel in a straight line. 'Our clothes were in rags,' wrote Galton, 'and at first our skins were very painful . . . especially as the scratches generally festered, but we got hard in time.' At one point he stopped to measure the strength of the thorns by tying a rag to a spring balance and pulling. One thorn withstood a pull of twenty-seven pounds.

With so many mouths to satisfy, water became a constant source of anxiety. Each day, after only three or four hours on the road, Galton would start to think about where to find it. Sometimes he would be guided to pre-existing wells. But usually wells had to be dug afresh, which could end up taking most of the day. Having found a good supply it was the norm to fill up before continuing the journey. On one occasion, however, Galton found himself without a suitable container. 'I could not think what to use as a water vessel, when my eye fell upon a useless cur of ours, that never watched, and only frightened game by running after them, and whose death I had long had in view.' Galton had learned that the skin of a dog is more watertight than most, so he simply killed the dog and turned its coat into a water bottle. Although the water kept Galton alive, he hadn't reckoned on canine justice. 'His death was avenged upon me in a striking manner', when, during the night, 'a pack of wild dogs came upon us, scattered our sheep . . . and killed them all.'

While water was a constant cause for concern, the diet was as predictable as the rising sun. Meat was the only thing on the menu, and each man ate about four pounds of it a day. Although meat was often plentiful, superstitions among the locals could make meal times a complicated affair. Different Damara families believed that they would be cursed if they ate oxen or sheep of certain colours. Since Galton had men in his command from many different families, all with their own specific beliefs, finding an animal that was acceptable to everyone was sometimes a challenge.

Keeping the crew fed and watered was not enough to prevent dissent among the ranks. But Galton was having none of it. Petty theft, quarrels, and occasional talk of mutiny turned him into a veritable Captain Bligh. His school days in Birmingham had evidently left their mark because he seemed to run his crew along similar lines. It was a rare day when he did not preside over one of his hastily arranged courts of 'justice'. Guilty was the usual verdict and the punishment a public whipping deftly administered by Hans Larsen.

The floggings may have helped maintain discipline but they did little for team morale. Three weeks and 150 miles after leaving Schmelen's Hope, the great lake of Omanbondè was still nowhere to be seen. Galton's men were entitled to feel a slight sense of trepidation. They were travelling through grim, war-ravaged country, to a destination whose existence was hard to corroborate. A current of desperation ran through the caravan.

Spirits were raised by a chance encounter with a couple of local Bushmen who seemed to know all about the lake. It was about four days' walk away, they said, and as broad as the heavens. At last, some good news. After the hardships of the previous month Galton and his men were looking forward to a lakeside camp, with some fishing and a spot of hippo hunting.

A couple of days later they bumped into a Damara man who offered to guide them the remainder of the way to the lake. On 5 April, the guide pointed out a low hill topped by a cluster of camelthorn trees. These were the tell-tale trees from which Omanbondè was supposed to have derived its name. A grassy river bed led up to the foot of the hill and then turned its corner. Behind and beyond, lay the cool blue waters of the lake.

Galton blazed a trail down the river bed, the wagons racing to keep up with him. But when he emerged on the other side of the hill there was no lake, only more dry river bed. He pressed on, hoping to catch sight of the lake around the next bend. But something wasn't right. The Damara guide had stopped behind the caravan and was looking around on the ground. Galton turned to confront the terrible truth. They had indeed arrived at their destination. They were standing right on top of it. This was the great lake of Omanbondè. And it was empty.

An exceptionally dry year had drained the lake of its water and the hippos had gone with it. The only damp thing in this remote part of Africa seemed to be Galton's spirit: 'I longed for the free and luxuriant vegetation of the tropics, and to emerge from a country that was scorched with tropical heat, but unrefreshed with truly tropical rains.'

With the lakeside holiday impossible it was pointless to hang around. They were now very close to the northern border of Damaraland. Beyond lay Ovampoland, a nation of advanced, agricultural people held in high regard by the Damara. According to local knowledge, the Ovampo king, Nangoro, was the fattest man in the world, and Galton was keen to meet him. So after a few days rest everyone was back in the saddle heading north again.

At the frontier of Damaraland the countryside changed abruptly. The thickets of thorn bushes that had made ribbons of skin and cloth were suddenly gone. In their place were dense timber forests, the nearest thing to jungle on the entire journey. While the tall trees offered some welcome shade from the burning sun, they were an obstacle course for the cumbersome wagons and a shattered front axle soon brought the whole caravan to a grinding halt. It would take weeks to fashion a repair. So Galton decided to continue on foot, taking a small party of men and oxen and leaving the rest of the caravan behind.

A large tract of mountainous terrain separated Ovampoland from Damaraland, and he needed a local navigator if he was ever going to find his way through the region. He'd detected reluctance on the part of the local Damara chiefs to provide him with a guide. Suspicions still surrounded Galton and his team. The Damara had no wish to upset the Ovampo people by dispatching a potential spy into their midst. Instead they suggested Galton should wait to join one of the Ovampo caravans of cattle traders from the north, which made fairly regular trips into Damaraland. But Galton had other ideas, and he managed to persuade a local Damara man to take him across the mountains. When it came to figuring out the route, however, the guide seemed remarkably forgetful. On the morning of the second day the guide confessed that he was completely lost. By the evening he had re-established his bearings by following a troop of baboons to a well-known watering hole. But the next day he was lost again, and Galton was beginning to lose his patience: 'A Damara who knew the road perfectly from A to B, and again from B to C, would have no idea of a straight cut from A to C: he has no map of the country in his mind, but an infinity of local details.'

Perhaps the guide was far more clever than Galton could ever conceive. If he was deliberately leading Galton up the garden path then his efforts paid off. A few days later, still stalled in the mountains, the touring party was intercepted by a group of twenty-four Ovampo traders heading south into Damaraland. The Ovampo captain, Chik, asked Galton if he was a rainmaker. 'I regretted that we were not, else we could travel when we liked and where we liked, and be independent of guides.' If Galton couldn't summon showers then the Ovampo insisted that he go back to his encampment with the wagons, and wait there until they had finished their trading. They would then guide him north into Ovampoland on their return.

Galton was impressed by the Ovampo. Their calm, ordered, and understated demeanour contrasted with the more volatile natures of the Damara people. Tall, slim, and muscular, they cut imposing figures:

> Their heads were shaved, and one front tooth was chipped out. They carried little light bows three and a half feet long, and a small and well made assegai in one hand. On their backs were quivers, each holding from ten to twenty well-barbed and poisoned arrows, and they carried a dagger-knife in a neat sheath, which was either fixed to a girdle round the waist, or else to a band that encircled the left arm above the elbow. Their necks were laden with necklaces for sale, and every man carried a long narrow smoothed pole over his shoulder, from either end of which hung . . . little baskets holding iron articles of exchange, packets of corn for their own eating, and water bags.

It took the Ovampo a couple of weeks to conclude their business in Damaraland, exchanging necklaces, beads, and small iron articles for over two hundred cattle. Then they were on the road again, heading back north. Joining them for their return journey were over one hundred Damara men and women, looking for husbands, trade or work in Ovampoland. And buried somewhere among the crowd were Galton and his team of thirteen men.

Halfway through the journey the caravan came across the sacred water hole of Otchikoto, over one hundred metres wide and surrounded by tall, vertical cliffs. Galton got out his plumb line to show that the lake sank to a depth of over fifty metres. The

Ovampo believed that this deep, bucket-shaped hole held mystical properties. No living thing that entered the water could come out alive. But Galton and Andersson chose to dispel the myth by stripping off and jumping in for a swim. The Ovampo had been very friendly up until that point, but they were disturbed by this apparent act of magic. From now on suspicion followed Galton's every move.

The hills and mountains eventually gave way to flatter country, but the environment remained harsh and hostile. On 30 May 1851, the caravan trudged its way through the baking atmosphere of the Etosha salt plain. Then they were back among the infernal thorns again. In the searing heat, Galton's destination seemed like a mirage, as elusive as the great lake of Omanbondè. Perhaps there was no Ovampo nation? Only a land of lacerating thorns.

And then the thorn bushes were gone, replaced by a scene of such contrasting fertility that it was difficult not to feel deceived. Cattle grazed in pastures alongside tidy fences that marked the neatly drawn boundaries of Ovampo homesteads. Fields of yellow corn, dotted with sumptuous palms, stretched out towards the horizon. It was a marvellous sight. After the agony of the thorns, Galton had stumbled across a little slice of paradise.

Not everything, however, was so rosy. His oxen were weak and nearing starvation, in desperate need of water and good grazing. But Chik, the Ovampo captain, seemed reluctant to offer more than the most meagre rations. Galton suspected that this was a deliberate ploy. With his cattle in poor condition it would be difficult for him to go wandering off on his own. Not that there was much chance of that. With no obvious landmarks in the vast, undulating sea of corn there was no way he was ever going to find his way without a guide.

The caravan travelled on for a couple of days, moving along narrow pathways through the corn stubble. At a clump of shady trees Chik ordered Galton and his men to set up camp. King

Nangoro lived barely a mile away and they were told to wait for his arrival. Galton still felt uneasy. The locals seemed polite but circumspect and they all kept their distance. Chik explained that everyone would relax once Nangoro had given his formal approval to their visit.

It was only natural that strangers were treated with a certain degree of suspicion. But Galton and the other Europeans were no ordinary strangers; they were probably the first white people the Ovampo had ever seen. Although commerce had been established with Portuguese traders to the north, these people must have been black, because the Ovampo showed a continual fascination with Galton's straight hair and white skin. 'They wondered if we were white all over,' wrote Galton, 'and I victimised John Allen [Hans Larsen's sidekick], who had to strip very frequently to satisfy the inquisitiveness of our hosts.' The Ovampo had difficulty imagining a country populated by whites alone, and were convinced that Galton and his Caucasian friends were rare migratory animals

A camp in Ovampoland

'of unaccountable manners but considerable intelligence, who were found here and there, but who existed in no place as lords of the land'.

While he waited for the arrival of the King, Galton prepared his collection of regal gifts. In truth, it amounted to a rather sad and pathetic offering. The gift of an ox was *de rigueur* in these parts and was his only present of any substance. The rest was mainly cheap rubbish that he'd picked up from costume stores in the West End of London: 'My position was that of a traveller in Europe, who had nothing to pay his hotel bill with but a box full of cowries and Damara sandals. I would have given anything for ten pounds' worth of the right sort of beads.' He had already made his first major *faux pas* by not sending Nangoro a present in advance. Now he feared that his paltry mix of gifts would add insult to injury.

The next day Chik came running to tell Galton that the King was on his way. Galton just had time to tidy up and organise his things before the entourage arrived: 'There was a body of men walking towards us, and in the middle of them an amazingly fat old fellow laboured along; he was very short of breath, and had hardly anything on his person.' Galton's gracious bow meant nothing to the King, and there was an uncomfortable pause as neither man seemed sure of what to do next. Wounded by the King's apparent indifference, Galton sat down and self-consciously added some footnotes to his journal.

After a few minutes the King gave Galton a gentle poke in the ribs with his walking stick and then joined him on the floor for a chat. He was accompanied by three well-dressed courtiers, who waited on his every move. They 'laughed immoderately whenever he said anything funny,' Galton recalled, 'and looked grave and respectful whenever he uttered anything wise.'

Once the formal introductions were finished it was time for Galton to get out the presents. As he handed them over he could only watch helplessly as the King's bucolic smile turned into a mournful frown. Galton was right. Gilded bits of tat just didn't cut it in these parts, or anywhere else for that matter, and the King wasn't shy in showing his feelings. He sank into a deep sulk. The courtiers dutifully followed suit.

Galton's poor gift selection was just the first in a succession of cultural blunders. The Ovampo believed that all strangers had the magical power to charm a life away. To protect themselves from this mortal curse, a counter charm was used, developed by Nangoro himself. Just before a meal, a mouthful of water would be spat out over the seated stranger's face. It was simple and, apparently, effective, but Galton balked at the idea of an Ovampo face wash and was having none of it.

Galton's obstinacy hardly engendered an atmosphere of trust between himself and Nangoro. But his biggest gaffe was yet to come. When he turned in for the night, he stepped into his tent to find a beautiful young woman dressed in little but beads and rings. She had smeared herself in butter and red ochre, and the golden glow of an oil lamp sent flickers of light across her breasts and thighs. Her open and affectionate face suggested a night of passion.

In an attempt to patch up relations, the King had presented Galton with his niece Chipanga as a temporary wife for the night.

Galton's illustration of the crowned King Nangoro

But while Chipanga did her best to get him in the mood, a nervous Galton could only fret about getting greasy fingerprints all over his white linen suit. She was, he claimed, 'as capable of leaving a mark on anything she touched as a well-inked printer's roller . . . so I had her ejected with scant ceremony'.

The rebuke was a personal insult to both Nangoro and his niece. Galton tried to repair some of the damage by making a big show of the crown he'd bought from the theatrical store in Drury Lane. But while the King and his courtiers seemed charmed by the ceremonial headdress, overall relations remained sour. When Galton brought up the idea of taking his expedition further north towards the Cunene River, Nangoro was discouraging and made no offer of assistance.

Galton's inability to get his own way in Ovampoland contrasted markedly with his experience in other parts of Africa. Amid the feudal fiefdoms of Damaraland he could strut about the place like a mini-dictator, doing pretty much what he wanted. But Ovampoland was a much more organised and sophisticated society, with a single king presiding over a strict system of rules. Guests – especially uninvited ones – were naturally expected to conform to local customs. Galton's intransigence only alienated him from the locals.

Nangoro started dropping heavy hints that Galton's departure south might be in the best interests of everyone. He certainly wasn't offering any encouragement to stay. Galton's oxen were continually denied the good grazing they needed. All they could hope for was a piece of ground 'as barren as Greenwich Park in summer-time', and they returned from the fields each evening a little thinner than they had been the day before.

With the Cunene River no more than four or five days' journey away Galton wasn't keen to leave. He had heard stories of Portuguese traders operating north of the river. To link up with this main waterway would make a satisfactory conclusion to his exploration. But after two frustrating weeks he agreed, reluctantly, to abandon his plans. And while there was great disappointment in having to head back, there was, at least, a consolation in the return of familiar freedoms: 'It was with the greatest relief that I once again felt myself my own master, and could go when I liked and as I liked; anything for liberty, even though among the thorn bushes.'

The 500-mile return journey back to Barmen was long but uneventful. There was the usual day-to-day drudgery of digging out wells, but by now Galton had become something of an expert at detecting water sources, and there was far less anxiety than had accompanied the outward trip. At the Damaraland border Galton picked up Hans Larsen and the rest of the men with the repaired wagons, and the completed caravan headed south.

On 4 August 1851, seven weeks after setting out from Nangoro's village, Galton was back in Barmen. 'I have just returned', he wrote to Frederick Campbell, 'to the most advanced missionary stations after my exploring journey, which indeed led me through a country most desolate, thorny and uninteresting.'

> But the end of it quite repaid my trouble, for I came to a peculiarly well civilised (if I may use such a word) nation of blacks where I was received most kindly but beyond whose territory I was not permitted to pass . . . Still I consider that I have completed the road from the Portuguese boundaries to the Cape, for the small intervening tract of land which I have not seen is well inhabited and well watered. My furthest point was Lat. 17° 58', Long. 17° 45'. The nation I reached was the Ovampos, governed by a fat stern king. I crowned him with all solemnity. His country is most fertile . . . They have more than one sort of corn; that which they prize the most is, I believe, unknown in Europe; it is certainly unknown in the North and the East of Africa.

Galton still had plenty of time on his hands. The next ship in Walfisch Bay was not expected until December at the earliest, at least three months away. Within a couple of days he hatched a plan to head east, towards Lake Ngami, to see if a simple route existed between the coast and the interior. Although this second excursion took the guise of genuine exploration, Galton was well aware that big-game country lay out to the east. If he could open up a new route to the interior then that would be good. If he could kill some game along the way then that would be even better.

On 13 August, less than ten days after his return from Ovampoland, Galton was on the road again. On his way east he dropped in on Jonker Afrikaner to pay his respects and commend him on keeping the peace during his absence. The region had been relatively calm since Galton's scene-stealing performance a few months earlier and, on the surface at least, it seemed that relations between the Damara and Nama people were improving.

Elephant Fountain, 200 miles east of Barmen, was the furthest east that the Namibian-based missionaries had penetrated. The spring was a meeting point for herds of zebra and roebuck, drawn in from the arid plains to drink. It was also the place where Galton teamed up with Amiral, a local Nama chief, and forty of his men. Together they would form a formidable shooting party.

There was game to be had around Elephant Fountain, but the animals for which the spring was named had long since disappeared. Galton wanted the biggest game that Africa could offer. But to find elephant and white rhino he and Amiral would have to travel another 150 miles inland to Tounobis, the mother of all shooting spots.

Even in the dry season water was abundant in the limestone rock pools that punctured the river bed at Tounobis. For wildebeest, zebra, rhino, elephant, and countless other species, this was the place to come after a long day's grazing on the plain, a comfy spot to cool down and quench your thirst, a place, perhaps, to find some relief from the day-to-day anxieties of savannah life. But it was not a place to come when Galton and his African acquaintances were in town. Dozens of animals were massacred in the killing frenzy that followed their arrival. Some were simply shot, others stumbled to their deaths in vast pitfall traps dug out of the ground. On occasions the scene resembled nothing more than a vast open-air butcher's. Rhino meat turned out to be a particular favourite of Galton's, the younger the better: 'A young calf, rolled up in a piece of spare hide, and baked in the earth is excellent.' The slaughter went on for days, but after a week even Galton felt he'd had enough.

Lake Ngami was still 200 miles to the north-east, but Galton conceded that it was a stage too far. Time was running out and his cattle were getting weak. Conversations with local tribesmen confirmed that the journey from Tounobis to the lake would be a difficult but doable one for the wagons. With this information Galton was able to take home the knowledge that a wagon route did exist from the coast to the interior. Content with his achievements, he packed up his oxen and left the killing fields behind. He was on his way back to Walfisch Bay for the very last time.

He arrived at the Bay in early December and had to sit out a frustrating month. 'This note I send by a ship now in the Bay and I wish much that I could go with her,' he wrote to his mother.

> But I have to look after my men . . . It is now nearly two years since I have heard anything of any description whatever from home, so that I am getting very anxious for my letters . . . St Helena is now my first point, it may be even 3 months before I am there, though I hope it will be much sooner.

When his ship finally arrived in January, he raced out in a small boat to meet it. Back on board, alone in his bunk, he listened to the gentle lullaby of waves against timbers. It was a soothing sound that seemed to drain all the worries from his body. He had forgotten what it was like to be free from the burden of African exploration – free from the sand that greedily swallowed every aching footstep, the cattle that refused to be tamed; the strange and unfamiliar languages of native tribes; the warlords that had to be appeased; the theft and mutiny; the difficulties of finding your way; the infernal thorns; the hunger and fear of starvation; and the endless worries about water. All that was over now, disappearing fast in the distant fog of Walfisch Bay. At last he could finally throw his anxieties overboard and sink into a sound and secure sleep.

6

A Compendium for Crusoe

It is better to think of a return to civilisation, not as an end to hardship and a haven from ill, but as a close to a pleasant and adventurous life.
Francis Galton

Galton wasn't the first European to make a name for himself as an African explorer. But he did perfect a style of exploration that was uniquely his own. He returned to England with a vast library of numbers: long lists of longitudes, latitudes and altitudes, the raw material that would give the mountains, valleys, and lakes of the Namibian landscape a more vivid, three-dimensional reality. With his astonishing attention to detail, he redefined what it meant to be a meticulous cartographer, setting new standards for future exploratory geographers to follow.

The Royal Geographical Society was highly impressed by Galton's fact-finding mission to the Namibian interior. His twenty-page report to the Society was a remarkably clear and informative account of the route he had taken, and the people and places he had encountered along the way. He provided neat summaries of the travelling conditions, the vegetation, landscape, and climate. He highlighted the areas where water was abundant and where it was scarce. He talked about Namibian history, demographics, and ethnic conflict. And when he had finished describing all he had seen for himself, he went on to discuss what he was told about the people and places that lay beyond his reach.

In recognition of his achievements, the Royal Geographical Society decided to honour him with one of their two annual gold medals, awarded for 'the encouragement and promotion of geographical science and discovery'. There was no doubting the prestige of the award. David Livingstone had won the medal in 1850 for his discovery of Lake Ngami, and future winners would include such distinguished names as Alfred Russell Wallace, Captain Robert Scott, Edmund Hillary, and Thor Heyerdahl. The award ceremony took place at the Society headquarters in Savile Row, and *The Times* sent a reporter to cover the event. In front of a packed audience it was left to the Society President, Sir Roderick Murchison, to make the presentation. After a short introduction, he turned to address Galton: 'It is now my pleasing duty to place in your hands this testimony of the approbation of the Royal Geographical Society.'

I am sure you will receive it, as we intend it, as the highest honour which we can possibly confer. You left a happy home to visit a country never before penetrated by a civilised being. You have accomplished that which every geographer in this room must feel is of eminent advantage to the science in which we take so deep an interest. Accept, with these expressions, my belief that, so long as England possesses travellers with the resolution you have displayed, and so long as private gentlemen will devote themselves to accomplish what you have achieved, we shall always be able to boast that this country produces the best geographers of the day.

For Galton, the medal was a royal stamp of approval on a budding scientific career. It cemented his reputation in the eyes not only of the scientific establishment, but the public at large. Once the newspapers had picked up on his adventure, the whole world knew about Francis Galton.

Galton was uncomfortable in the role of national celebrity, and he retreated from public view in the only way he knew how, heading off to Norway for a summer's sailing and fishing with his old hunting friend, Sir Hyde Parker. While the holiday kept him out of the spotlight it did little to improve his ailing health. An intermittent fever was one of the less coveted souvenirs he had brought back from Africa. He tried to shrug off the illness, but by the autumn of 1852 it was becoming a serious cause for concern.

Ignoring the advice of friends and family, he insisted on travelling to London to attend the funeral of the year. Wellington, the Iron Duke, was dead. His passing was a symbolic moment in British history, a point when the country consciously let go of its past. The warrior who had defeated Napoleon on the battlefield at Waterloo represented a bygone age; his passions and his politics had fallen out of step with the new climate of diplomacy that now characterised Victoria's reign.

Galton was keen to join the crowds and pay homage. He had been away in Africa for the previous year's big event, the Great Exhibition of 1851, and he was determined not to miss out this time. But the occasion took its toll on his fragile health. Unable to find a seat, he was forced to stand in the cold for hours as he waited for the funeral cortege to go past. By the end of the day he was exhausted. Near collapse, he struggled home to his mother's country estate in Claverdon, and sat out the rest of the year to recuperate.

At Christmas he joined his mother and sister Emma for a holiday in Dover. The British seaside was putting on its classic winter display. Dover was damp and dull, a melancholy mix of browns and greys. The sun disappeared for days on end, drowned out in the swell of cloudy gloom. Out on the promenade ageing couples made the most of what light was available, snuggling up on sea-front benches, anchored down against gale-force winds and horizontal rain.

Yet the dreariness of Dover turned out to be just the tonic that Galton needed. The billowing sea air seemed to inflate his wilted body and bring him back to life. Suddenly he was up and about and eager to socialise. Over the Christmas and New Year period he threw himself into a string of social engagements, turning up at dinner parties and society balls, happy to play the part of the celebrated African explorer.

Galton's renaissance was about more than just the end of an illness. It also signalled that he was emerging from a period of intense personal grief. Nine months earlier, on his way home from Africa, he had learned that his great friend Harry Hallam had died at the age of twenty-six. Hallam had been a compassionate confidant in his bleakest Cambridge days, someone with whom he had walked and talked his way through much of his idle twenties, and his death left him distraught. Back in Britain he had made a pilgrimage to the memorial where Hallam had been laid to rest alongside other family members, and, 'kneeling alone in a pew by their side, spent the greater part of a solitary hour in unrestrained tears'.

Hallam's death had taken some of the gloss off Galton's celebrated return to British shores. It had also left him with a large hole in his life where a great friend used to be. Whether Galton had this in mind as he embarked on his south-coast social revival is anybody's guess. But, now in his thirties and aged even more by his illness, this available bachelor allowed himself to think about marriage.

It was at a Twelfth Night party in Dover that Galton met his future wife, Louisa Butler. Louisa hailed from a distinguished academic family. Her father was not only an ex-Cambridge man, but a senior wrangler, the most eminent mathematician of his year. He was headmaster of Harrow before assuming his position as dean of Peterborough cathedral. As well as his academic abilities, 'he had been noted for athletic powers, and he much prized a medal awarded to him by the Humane Society for having saved the life of a drowning woman when long past his middle age'. Louisa's four brothers were also top-class university men. The youngest, a senior classic at Cambridge, went on to become Master of Trinity, Galton's old college. The others reached the pinnacle of their chosen careers, be it in law, education, or government administration. All in all, the Butlers were an imposing bunch of outstanding achievers, and a focal point for high society.

From the simple and rather limited perspective of photographic evidence, it is difficult to see what attracted Galton to Louisa. Physically she was unremarkable, conservative in both dress and appearance. Her countenance suggested a solemn and melancholy disposition. She had a round face with a low-set mouth, a large beaky nose and rather hollow, empty eyes. Galton himself was not classically handsome, yet there was still something striking about his appearance. He was slim and physically fit, and at five feet ten, he stood relatively tall for the times. His blue eyes were tucked away beneath a heavy brow that prompted comparisons with Charles Darwin. Like his half-cousin, Galton was balding by his early twenties, but he made up for the loss on top with a pair of extravagant sideburns.

If physical beauty could not account for Louisa's appeal, then perhaps her character and personality could. But here again, the details didn't add up. Although both of them were politically and socially conservative, Louisa was a lover of music and fine arts, two things guaranteed to send Galton running for cover. Despite dabbling with opera, he had discovered his own musical limitations at Cambridge: 'I find the Galton ear is as slightly developed inside my skull as it is largely on the outside, and although I keep up the credit of the family failing, yet I am afraid I shall not at the same time qualify for the professorship of music.'

Religion was another significant point of difference. Louisa was a devout Anglican. Galton, on the other hand, was far less constrained. His Quaker ancestry had given him a more pragmatic approach to religion, and although a spiritual side to his character had been revealed at Cambridge, he was beginning to err on the side of agnosticism.

On the face of it, there seemed to be little common ground

between Galton and Louisa. They were both the same age – thirty – but shared little else. Galton devoted chapter eleven of his auto-biography to the subject of his marriage. It is one of the shortest chapters in the book and contains no reference to where or when he first met his wife, nor any mention of the wedding itself. We learn nothing of Louisa's appearance, age, or interests. In fact he didn't even give his wife's name. Perhaps he was too shy or too modest to highlight his affection for Louisa. Or perhaps not. After all, this is the same book in which we get to hear all about the 'singular sweetness' of Harry Hallam. What we do have, however, in these nine brief pages, is dense detail on who, among Louisa's many eminent relatives, won what scholarship to where, who was master of which Cambridge college, and who was high up in the Indian Administration. To judge by this account Galton's attrac-tion to Louisa was motivated not so much by the woman herself, but by the intellectual friends and relatives she brought with her, a fact he was not frightened to spell out: 'I protest against the opinions of those sentimental people who think that marriage concerns only the two principals; it has in reality the wider effect of an alliance between each of them and a new family.'

Once Galton had locked onto the idea of Louisa, he was forth-right in demanding her hand in marriage. He left Dover in March 1853, and proposed the following month. By the end of April he had obtained formal consent from her father, and everything looked set for a hastily arranged spring wedding. But his marriage plans were about to suffer a serious setback. On Saturday 30 April he set off to see his fiancée and future father-in-law at their home in Peterborough, and was in high spirits when he arrived that evening. But instead of bursting in on a scene of family bliss, he walked in on a wake. George Butler, Louisa's father, had died of heart failure that afternoon.

The death of her father at this key moment in her life may have brought Louisa closer to her fiancé. But it's impossible to know what Louisa was thinking. There is no record of her thoughts and feelings at the time. One of the few pieces of correspondence still surviving from this period is from Louisa's sister Emily. It was written in May 1853, barely a month after her father's death. From this it's clear that Galton was perceived by Louisa's family as a dashing and heroic adventurer:

> Mr G. came yesterday fresh from the Derby; I felt so pleased to have such a sportive relation. It was a splendid day at Epsom, and he was very happy wandering among the gypsies etc. He tells such rich stories and very neatly. He has been to spirit rappings [seances] and had another conversation in Damara with a deceased chief of that tribe. Is not that wonderful, for Mr Galton is the only man in Europe who knows Damara. The chief promised to go abroad with him, which is a pleasant look-out for Loui!

Galton and Louisa were eventually married on 1 August 1853. They spent the rest of the year honeymooning in Italy, Switzerland and France. Long tours, sometimes at home but usually abroad, would become a feature of their future life together. They frequented stylish resorts and spa towns: Vichy, Wildbad, Baden and Mont Dore les Bains would become some of their favourite haunts. 'Certainly,' Galton wrote in his autobiography, 'we led a life that many in our social rank might envy.'

Love, real or illusory, seemed to suit Galton. In early 1853 the illness he had brought home from Africa abated, and he was full of energy and ideas. With the backing of the publisher John Murray, he had begun work on a full-length popular account of his travels in Namibia. He dashed off the book in a couple of months, submitting the 200-page manuscript in late April. The result, *Tropical South Africa*, was released on an unsuspecting public later that year.

Those who ventured to buy the book – and there were quite a few – were rewarded with a classic Boys'-Own adventure, written in a unique style. The book, like the journey itself, is full of peculiar

The happy couple: newly-weds Francis Galton and Louisa Butler

experiences, and it is difficult to pick out a typical passage. But the following extract offers a flavour of what lay inside. Galton and the caravan were on their way to the mission station at Otjimbinguè. They had just lost a mule and a horse to the hungry lions of the Swakop basin, and now their own provisions were perilously low. Desperate for food, Galton had picked up the trail of a lone giraffe.

After a few hours' travel, during which I had kept a couple of miles in front of the rest of the party, so as to be well away from the sound of the whip and of the men's talking, the tracks turned sharp to the right, up a broad ascent, which there led out of the river, and in the middle of this, amongst some bushes, and under a camelthorn tree stood my first giraffe. I took immediate advantage of a bush, and galloped under its cover as hard as I could pelt, and was within one hundred yards before the animal was fairly off. I galloped on, but she was almost as fast as I, and the bushes, which she trampled cleverly through, annoyed my horse extremely; I therefore reined up, and gave her a bullet in her quarter, which handicapped her heavily, and took some three miles an hour out of her speed. Again I galloped, loading as I went, but excessively embarrassed by the bushes, and fired again, whilst galloping, at thirty yards' distance, and I believe missed the animal.

With his horse tiring, Galton struggled to keep up with the wounded giraffe.

At last I steadily gained on the giraffe, then beat her, and passed her. The giraffe obstinately made for her point. I was forty yards in advance, and pulled up full in her path. She came on: my horse was far too blown to fidget, and was standing with his four legs well out. I waited as long as I dare – too long, I think, for her head was almost above me when I fired, and she really seemed coming at me with vice. I put my bullet full in her face; she tossed her head back, and the blood streamed from her nostril as she turned and staggered, slowly retracing her path. I dare not fire again, lest I should fail in killing her, and only excite her to another run, which my horse was not fit to engage in. I therefore rode slowly after the wounded beast, and I drove her back to near where she came from, and there she stopped under a high tree. My horse was now frightened, and would not let me take my aim for the finishing blow at the brain, as it is but a small mark to shoot at; so I got off, and the unhappy creature looked down at me with her large lustrous eyes, and I felt that I was committing a kind of murder, but for all that, I was hungry, and she must die; so I waited till she turned her head, and then dropped her with a shot.

Galton's writing had a stream-of-consciousness quality about it that could be either imaginative and charming, or disordered and annoying. Also in evidence was a remarkable attention to detail. He never allowed the reader to forget how many men, oxen, sheep, and horses were with him at any one point in his journey, even if it meant confusing the course of events. Nevertheless, the overall impression was startling. It was impossible not to sit back and admire the courage, endurance, and, at times, sheer madness of the whole exercise, and Galton told it all with a rare combination of deadpan wit and modesty, as if he was describing nothing more than an average day at the office.

There was one other aspect to *Tropical South Africa*, however, that revealed an altogether different side of Francis Galton. Throughout the book Galton airs his views on the local African people that he meets on his journey. These are not carefully considered anthropological insights, or deft illustrations of cultural contrasts. They are vicious, racist rhetoric. His first meeting with the Nama tribesmen at Walfisch Bay sets the tone: 'Some had trousers, some coats of skin, and they clicked, and howled, and chattered, and behaved like baboons . . . I looked on these fellows as a sort of link to civilisation.' The Nama got off lightly compared to the Damara people. Galton provided an ongoing ridicule of the Damara's character, customs, and culture, embodied in frequent, curt summations: 'There is hardly a particle of romance, or affection, or poetry, in their character or creed; but they are a greedy, heartless, silly set of savages.' Only the Ovampo people, with their more recognisable style of monarchy and social order, were singled out for any kind of praise.

To a twenty-first-century reader, these passages stand out like sores on the page. To Galton's Victorian readers, however, the issue would have barely registered, if at all. Even though Britain had abolished slavery in 1833, Galton's assessment of African peoples as generally uncultured and inferior merely echoed the racist consensus then prevalent throughout Europe and the United States.

Although racist attitudes were not new to the nineteenth century, the rationale behind the racism had shifted significantly since the days of the Enlightenment. Race, in the Victorian sense of the word, was a broad term that represented differences not just in ethnicity, but in class and social status. In the first decades of the nineteenth century many European thinkers had begun to question why social divisions remained so entrenched in prosperous and progressive nations like Britain and France. Instead of emancipating the people, the industrial revolution had merely reinforced social hierarchies and, in some cases, created new ones. The inner-city slum, for instance, became a potent image of the early Victorian era and a source of great social and philosophical concern. Why did such grinding poverty persist? Why had progress touched the lives of some people and not others?

The Victorian solution to this problem was to move away from the Enlightenment view that all men were inherently equal, and to assert the exact opposite. In this revised social scheme, the poor could not escape their poverty because they were intrinsically incapable of doing so. Social hierarchies, in other words, were seen as a reflection of an underlying natural order. Those at the bottom were not only socially inferior, they were naturally inferior as well.

This view was extended to people of other nations. Why had the tribal civilisations of Africa not risen to the sophisticated standards set by Europeans? Because they were innately inferior was the answer. This was not the view of extremists, it was firmly embedded in the Victorian mindset. Consider this statement on Negroes, for instance, from Thomas Henry Huxley, one of the most progressive and liberal thinkers of his day: 'It is simply incredible that, when all his disabilities are removed, and our prognathous relative has a fair field and no favour, as well as no oppressor, he will be able to compete successfully with his bigger-brained and smaller-jawed rival, in a contest which is to be carried out by thought and not by bites.' Throughout the middle of the nineteenth century similar sentiments were expressed in the biology of Darwin, the politics of Hegel, the literature of Dickens, and the poetry of Tennyson.

Although you could argue a cultural or historical defence for Galton's comments, there still seems something startling about the racism of *Tropical South Africa*. The brusque arrogance with which he dispatches his assessments, and the sheer frequency with which he reiterates his racist views, suggests a hint of mania, a deep-seated passion in his prejudice. With the benefit of hindsight, these bleak, depressing passages read like ugly portents for his future.

Far from damaging his reputation the publication of *Tropical South Africa* seemed like an excellent career move for Galton. The Royal Geographical Society was so impressed that, in 1854, it elected him to their Council, to sit alongside such eminent geographers as George Everest, Surveyor General of India and the man who gave his name to the world's highest mountain, and Francis Beaufort, of wind-speed fame. The book's publication also drew him into contact with old acquaintances. Charles Darwin had lost touch with his younger cousin, but having read his book, felt compelled to write and congratulate him:

> I last night finished your volume with such lively interest, that I cannot resist the temptation of expressing my admiration at your expedition, and at the capital account you have published of it. I have no doubt you have received praise, from so many good judges that you will hardly care to hear from me, how very much I admire the spirit and style of your book. What labours and dangers you have gone through: I can hardly fancy how you can have survived them, for you did not

formerly look very strong, but you must be as tough as one of your own African waggons!

If you are inclined at any time to send me a line, I should very much like to hear what your future plans are, and where you intend to settle. I so very seldom leave home, owing to my weakened health (though in appearance a strong man) that I had hardly a chance of seeing you in London, though I have often heard of you from members of the Geographical Society.

I live at a village called Down near Farnborough in Kent, and employ myself in Zoology; but the objects of my study are very small fry, and to a man accustomed to rhinoceroses and lions, would appear infinitely insignificant.

The accolades continued to pour in. In 1854 the French Geographical Society bestowed upon Galton one of their highest honours when they awarded him their silver medal. That same year he was allowed to bypass a sixteen-year waiting list to gain membership of the exclusive Athenaeum Club on London's Pall Mall. This, perhaps, was the most satisfying honour of all. Galton had been dreaming of the Athenaeum since his days as a medical student at King's College. The club was a haven for the Establishment elite, a place where prominent politicians, writers and scientists could escape the pressures of public life, for quiet conversation over expensive brandy and Cuban cigars.

Despite all the plaudits, there was no let-up in Galton's work rate. His next publication, 'Hints to Travellers', appeared as part of a special edition of the Royal Geographical Society's journal in 1854. The article was essentially a detailed list and description of equipment that no intrepid explorer should leave home without. This simple concept proved remarkably popular. 'Hints to Travellers' was later published as a small book. It ran to several editions and became the Royal Geographical Society's most popular publication.

Within months he was expanding the 'Hints to Travellers'

theme into a much grander idea. From his own experience of travelling in hostile and unfamiliar terrain, Galton knew all about the importance of basic survival skills. Time and again in Namibia, simple bush lore had saved him from certain death. He realised how much smoother his own journey would have been had he had access to a comprehensive compendium of survival information. But no such book existed, and Galton felt that it was about time that he compiled one. This was not going to be a book for Sunday afternoon strollers in the Lake District. It was going to be a lifeline for the more adventurous traveller, small enough to fit into a jacket pocket but large enough to provide toilet paper in an emergency.

The Art of Travel appeared in 1855 and was an instant bestseller. It appealed not only to the genuine traveller, but to anyone with a morbid fascination for the discomfort and hardship of others. Galton had opened a window on the harsh and often brutal world of survival. To survive in the wilderness he showed that there was no choice but to join with nature, red in tooth and claw.

Galton provided novel and sometimes eccentric solutions to all kinds of potential problems. Imagine, for instance, that you were dying of hunger in the African bush and vultures were the only meat for miles. How on earth do you go about catching them? Turn to the section on hunting and Galton had the answer.

> Condors and Vultures are caught by spreading a raw ox-hide, under which a man creeps, with a piece of string in his hand, while one or two other men are posted in ambush close by, to give assistance at the proper moment. When the bird flies down upon the bait, his legs are seized by the man underneath the skin, and are tied within it, as in a bag. All his flapping is then useless; he cannot do mischief with his claws, and he is easily overpowered.

Of course the hunter can easily become the hunted. So what do you do when your touring party is shadowed by a hungry pack of wolves? Consider the following cunning solution from Scandinavia: 'The Swedes put fulminating-powder in a raw shankbone, and throw it down to the wolves; when one of these gnaws and crunches it, it blows his head to atoms.'

In the section on caches and depots Galton had some ingenious, if painful advice on what to do when visiting a 'rich but imperfectly civilised country'. Travellers who fear being robbed of all their clothes and valuables 'sometimes buy jewels and bury them in their flesh', he wrote. 'They make a gash, put the jewels in, and allow the flesh to grow over them as it would over a bullet . . . The best place for burying them is in the left arm, at the spot chosen for vaccination. A traveller who was thus provided would always have a small capital to fall back upon, though robbed of everything he wore.'

Among the more predictable passages on how to make fire, build shelters and find water, Galton sprinkled some genuine curiosities. In the clothing section, for instance, there was an instructive little paragraph on the best way of rolling up shirt sleeves. 'Recollect that the way of doing so is, not to begin by turning the cuffs inside out, but outside in . . . In the one case, the sleeves will remain tucked up for hours without being touched; in the other, they become loose every five minutes.'

Not all the information was so insightful. Galton's advice on how to turn your boat into a bigger one – 'saw it in half and lengthen it' – seemed to be asking for trouble. Some of his own theoretical contributions could also confuse rather than help matters. It was probably not a good idea, for example, to get lost on your journey. Otherwise you were faced not only with the risk of starvation, but five pages of Galton's mathematical theory on the best way to get out. These, however, were minor issues. Overall the book was a brilliant distillation of practical survival skills, exhaustive in both scope and detail. Not surprisingly, it became a seminal work for survivalists. Much of the information it contains is still relevant today and, after eight subsequent editions, it is the only one of Galton's books to remain in print.

Illustration from *The Art of Travel*, showing Galton's recommended technique for
steering a horse across water

Although *The Art of Travel* was written with explorers and
travellers in mind, Galton recognised that there were others
who would benefit from his advice. In 1855 the British Army
was hopelessly ill-prepared for its bloody Crimean campaign
against the Russians. While many held incompetent commanders
responsible for the thousands of British deaths, it was a complete
ignorance of camp life, as much as poor leadership, that was to
blame.

In the spring of 1855 Galton wrote to the War Office, offering to
give free lectures on basic survival training at the newly founded
military barracks in Aldershot. Given the air of confusion then
permeating the corridors of Whitehall it is no surprise, perhaps,
to discover that his letters met with a stony silence. So Galton
redrafted his message to the Prime Minister, Lord Palmerston, and
this time his offer fell on more receptive ears. Within weeks he was
installed at a cottage in Aldershot with all facilities at his disposal.

Galton's lecture course was a do-it-yourself guide to survival in
the wilderness. There were lessons on finding, purifying and

filtering water; making fires; building shelters; pitching tents; cooking and preparing food; tracking; hunting; and fishing. He gave instruction on the best way to fell trees, and on how to build bridges, boats, and rafts. He discussed how to make candles, soap, and other household artefacts, all from simple animal products. And he gave illustrative talks on hobbling horses, harnessing cattle, packing saddles, and loading wagons. Characteristically, the syllabus was extensive and detailed. Galton was leaving nothing to chance. At the end of the course a soldier should be able to wake up naked in the middle of nowhere and not only survive, but find his way back to civilisation.

The lectures attracted the interest of *The Times*, who sent one of their reporters to the Aldershot barracks to sit in with Galton's grunts. The subsequent report gave his course a deservedly positive write-up:

> The camp of Aldershot at this moment is the scene of a remarkable experiment. Mr Galton, a gentleman of considerable experience in the shifts and contrivances available for travelling in wild countries, has obtained the permission of Lords Panmure and Hardinge to tenant two huts with an enclosure adjoining for the purposes of communicating his experience to the British soldier. His services are rendered gratuitously at present, for his arts are untested and his teaching is a novelty. But the day will perhaps come when their value will be recognised and he himself be duly installed as our first Professor of Odology.

Galton gave three lectures a week from July to September 1855, and then came back the following spring to do it all over again. But while his aims were admirable, the military lacked enthusiasm, and Galton struggled to pull in the crowds. Lecture attendances varied from weak to woeful, with a maximum of fifteen and a minimum of three. There were mitigating factors to help explain the poor turnouts. Galton's lecturing style was dry and distant. It was all too evident that he lacked the instinctive ability to connect with the audience, and the audience seemed only too happy to reciprocate.

Galton's presentations may have been dull, but their content was designed to save soldiers' lives. Two years later, the military finally woke up to their value, and models, illustrations, and notes taken from Galton's lectures were distributed to military centres throughout Britain. Galton had made a small but significant dent on the philosophy of soldiering, and its impact was long-lasting. Today, traces of his art of travel can be found on every military curriculum.

The lacklustre response from the military men of Aldershot left Galton feeling dispirited, and he returned to London in the spring of 1856 unsure of where his life was heading. 'I was rather unsettled during a few years,' he wrote in his autobiography, 'wishing to undertake a fresh bit of geographical exploration, or even to establish myself in some colony; but I mistrusted my powers, for the health that had been much tried had not wholly recovered.' In the end he settled on a compromise. He would immerse himself in the activities of the Royal Geographical Society, living a vicarious life of exploration through the exploits of others.

The 1850s was an exciting time for exploratory geography. Richard Burton made his famous journey to Mecca and Medina disguised as a Moslem; Laurence Oliphant and Commodore Perry were opening up Japan to the outside world; John Rae of the Hudson's Bay Company was penetrating the North American Arctic, while closer to home, Ferdinand Lesseps was devising his grand scheme for a canal across the Suez isthmus. But the decade's perennial puzzle – and the ultimate prize of exploratory geography – was the source of the White Nile. Most agreed that the head of the river lay somewhere in East Africa, but there was little else to go on. Then, in the middle of the decade, East African missionaries began returning to Europe with tales of a vast snow-capped mountain, Kilimanjaro, and a great lake beyond. In 1856 a Royal Geographical Society-backed expedition, headed by Richard

Burton and John Speke, was sent out to Tanzania to investigate. Galton drafted the travel arrangements.

Together, Burton and Speke discovered Lake Tanganyika, after which Burton promptly fell ill, leaving Speke to go on alone and claim the prize discovery of Lake Victoria. But while the British public keenly followed the course of events in the east, Galton had half an eye on news emanating from his old stomping ground on the opposite side of the continent. Violence had broken out in the peace-loving nation of Ovampoland. The details of the skirmish were vague, but the outcome was unambiguous. King Nangoro, the fattest man in the world, was dead, shot down by European guns. The perpetrators were not Damara or Nama tribesmen, nor were they European bandits, slave traders, or cattle rustlers. They were European missionaries.

The missionaries had been part of an expedition retracing Galton's northern route to Ovampoland. Among the party were Mr Hahn and Mr Rath, missionary men with whom Galton had fraternized on his way east into the Namibian heartland. Though the Ovampo had initially welcomed the missionaries, an argument had developed over the correct way of giving gifts to the King, and the disagreement had somehow escalated into a full-blown battle. In the confusion that followed dozens of Ovampo were shot dead.

A report of the incident was read out at a meeting of the Royal Geographical Society, and Galton used the occasion to launch a savage attack on the missionaries, lambasting their inexperience and inability to conciliate. He concluded with a moving and impassioned defence of the Ovampo, lamenting what he considered were needless deaths: 'I was able to leave, in peace, the happy country of a noble and a kindly negro race, which has now, for the first time, been confronted and humbled before the arrogant strength of the white man.'

The Royal Geographical Society was quickly becoming Galton's second home. He was hardly a born bureaucrat, but a reforming zeal landed him in all kinds of administrative posts. Sidling onto various sub-committees, he was able to wield considerable influence over Society policy and decision making. He criticised the Society's annual journal, for instance, arguing that a more frequent publication was needed to keep abreast of all the latest geographical news and developments. He got his wish the following year when *Proceedings of the Royal Geographical Society* appeared in 1856.

Galton was also instrumental in getting geography onto the school curriculum. He had discussed the idea with the headmasters of a number of public schools, including his brother-in-law, George Butler, then master of Liverpool College, before persuading the Royal Geographical Society to put its weight behind the project. The result was an annual gold medal, awarded to the best geography student in the land. In practice, the award was only open to public schools, and only ten of these took any kind of active part. The medal was not an idea that won unanimous support among Council members, and there must have been suspicions of favouritism when, in only its second year, the award went to Galton's nephew, George Grey Butler. When some schools, most notably Dulwich and Liverpool, were accused of grooming students for the award just to gain prestige, the medal was quietly withdrawn.

In 1856, still only thirty-four, Galton was made a Fellow of the Royal Society. A year later, his active involvement in the Royal Geographical Society was rewarded with his appointment to Joint Honorary Secretary. He was juggling an increasing number of commitments, and there was no sign of any slowing down. In 1858 his expertise in the use of navigational and astronomical instruments earned him a position on the Kew Observatory Management Committee. The Observatory had made a name for itself as the place 'where the outfits of all magnetic observatories, English and foreign, were standardised, and where intending observers were instructed'. But by the time Galton joined, it had extended its remit to cover all kinds of scientific instruments, from thermometers and telescopes to watches and sextants. It was a natural environment for a man with an eye for detail.

Clubs and societies were great places to make new friends and acquaintances and, judging by his autobiography, Galton was making the most of his opportunities. In contrast to the chapter on his marriage, 'Social Life' is one of the longest chapters in the book. In it Galton names a long list of scientific greats, explorers, diplomats, and parliamentary figures with whom he wined, dined, walked, and weekended. By the end of the 1850s his social diary must have been bulging with eminent names.

Galton enjoyed a Cornish ramble with the physicist and mountaineer John Tyndall. The two men seemed well matched in character and temperament. Tyndall had calculated that the energy required to climb the Matterhorn in the Swiss Alps was the same as the energy contained in a ham sandwich. Although he failed on his first attempt at the mountain in 1860, he succeeded eight years later, sustained, so the story goes, on nothing more than a ham sandwich.

The botanists George Bentham and Joseph Hooker were fairly regular guests at the Galton household. Bentham was the nephew of Jeremy Bentham, the utilitarian philosopher and founder of University College London, and would himself become a celebrated president of the Linnean Society.

People, not plants, was the usual topic of conversation round at archaeologist John Lubbock's house. 'His week-end invitations were always most instructive and grateful,' remembered Galton. Lubbock was a kind of Charles Darwin of the archaeological world, the man who did more than most to outline the cultural and technological changes that took place in early human history. He and others mapped out the developmental path that led from the first primitive stone flints to the more accomplished tools of the iron and bronze ages.

It was on a weekend visit to Lubbock's home that Galton first met Herbert Spencer, the philosopher and social Darwinist who coined the term 'survival of the fittest'. Spencer interpreted

Lubbock's outline of technological progress as evidence of an underlying biological improvement, and he came to champion the popular mid-Victorian view that cultural and biological evolution were one and the same. Galton and Spencer were never close, although they did enjoy occasional days out at the Derby together.

Given the scientific circles in which Galton was mixing, it seems only natural that scientists should make up the majority of his friends. Even so, there was a conspicuous lack of literary figures within Galton's social sphere. In his autobiography the chapter on his social life contains the name of only one writer, Thomas Carlyle, but it is clear that they hardly considered each other a friend. Recalling one meeting Galton found Carlyle to be 'the greatest bore that a house could tolerate'. Carlyle 'raved about the degeneracy of the modern English without any facts in justification, and contributed nothing that I could find to the information or pleasure of the society'. Carlyle's ultimately pessimistic view of England and its future was evidently not to Galton's taste.

Fundamentally Galton was politically and socially conservative. While he could tolerate eccentricities in his friends, he was a traditionalist who paid due deference to establishment norms and values. He and Louisa were most at ease in the company of such official figures as John Crawfurd, the first Governor of Singapore, Frederick North, the member of parliament for Hastings, and the lawyer and Southampton MP, Russell Gurney. It was through his association with Gurney, perhaps, that Galton acquired his great reverence for the legal profession. Certainly, Galton could not speak of him more highly: 'I have known no one who struck me as a more just, searching, and yet kindly judge, or whom I would more willingly be tried by if I fell into trouble.'

Of the many new faces that Galton met during the 1850s, William Spottiswoode stood out from the crowd. Spottiswoode had energy coming out of his ears, and somehow managed to combine a distinguished day job as Printer to the Queen with a highly successful career in mathematics and science. He and Galton first met at the Royal Geographical Society in the 1850s, and Spottiswoode proved to be a valuable political ally in Galton's struggles with Society reform. At one time the two men had considered the possibility of a joint adventure to the Sinai peninsula. Galton even took Spottiswoode to the Isle of Wight for some preparatory instruction in surveying. 'We found a strongly railed field', recalled Galton, 'into which we got by climbing a fence, and prepared to unpack, not particularly noticing the cattle in it.' Out of the herd a bull emerged, advancing in 'so threatening and determined a manner that we had to retreat from the brute as best we could'. The Isle of Wight proved to be the limit of their exploration. After the raging-bull incident, Galton fell ill with a serious mouth abscess and Sinai was put to one side.

7

Storm Warnings

Those who are over eager and extremely active in mind must often possess brains that are more excitable and peculiar than is consistent with soundness.

Francis Galton

Rutland Gate is one of London's more affluent avenues, a broad South Kensington thoroughfare framed by neat terraces of white Georgian mansions, five stories high. Elevated porticoes rise from the street as if hinting at some higher plane of living. Strict colonnades of trees add to the sense of order while a gated park provides a tasteful reminder of the weekend retreat in the country. Today, the street groans under the weight of Bentleys and Jaguars, but not much else has changed since Galton and Louisa moved there 150 years ago.

The happy couple had spent the first four years of their married life living in rented accommodation in Holborn and Victoria. But once Galton had made his decision to abandon exploration and stay in Britain, it made sense to find a more permanent London home. In 1857 he paid £2,500 for 42 Rutland Gate. The central location offered a convenient base from which to entertain his growing circle of influential friends. The house was also ideally situated for easy access to the many clubs and societies where he was spending an increasing amount of his time. And Hyde Park was only a stone's throw away, a vast open space where Galton could enjoy his regular morning exercise.

Inside, the house's decor was bare and austere. Galton had a unique approach to interior design. His furniture was an unco-ordinated and eclectic mix of styles. Function, not form, was his primary concern and he was always looking for ways to make practical improvements. It annoyed him to climb the stairs, for instance, only to find the toilet occupied. So he installed a piece of frosted glass in the toilet door and added a vertical iron rod to the bolt. When the toilet was in use, and the bolt drawn across, the iron bar could be seen clearly through the glass from the bottom of the stairs.

The ground floor was dominated by a dining room where Galton would do most of his work, sitting at a desk in the front window. All his notes and manuscripts were kept in a store room at the back of the house, but there were precious few scientific books on display. Those books he had were usually complimentary copies given to him by friends and colleagues. In fact it is one of the more peculiar aspects of Galton's character that he rarely read the work of other scientists. When he approached a scientific problem for the first time he hardly ever bothered to research the published literature. Unburdened by history and the work of others, he tackled his subjects head on, from his own first principles. This unusual mode of working undoubtedly gave his research an original and fresh perspective that contributed to many of his successes. But it could also expose him to justifiable accusations of arrogance.

Sometimes there was very little published literature to consult. Galton was often at his most comfortable when labouring in uncharted scientific waters. In early 1859, for instance, he set sail on an experimental odyssey into the unknown. Making a tempo-rary move away from geography, he turned his thoughts instead towards that great British institution, the cup of tea.

Galton wanted to lay bare the art of tea making, to turn it into a theoretical science, and he set off in pursuit of the problem like a

man in search of a new universal law of nature. Twice a day for two
months the business of brewing came under severe scientific
scrutiny, as he endeavoured to discover what it was that made a
cup of tea taste the way it did.

Before he could begin he had to make some important modifications to his apparatus: 'I had a tin lid made to my teapot, a
short tube passed through the lid, and in the tube was a cork,
through a hole in which a thermometer was fitted, that enabled
me to learn the temperature of the water in the teapot at each
moment.' His kitchen became a laboratory in which the temperature of the teapot, the temperature and volume of the water,
and the weight of tea were all carefully measured and recorded.
Galton's personal preference was for tea that was 'full bodied, full
tasted, and in no way bitter or flat', but, in the interests of
scientific neutrality, he also garnered opinions from Louisa, his
servants and any friends and family members who happened to be
passing.

February saw a flurry of tea-making activity, and extracts from
his notebooks reveal the methodical steps he was making towards
cracking the brewing code. On the morning of Friday, 18 February he reported that the tea was quite good: 'I think it could
bear strengthening. LG [Louisa Galton] says not.' The next day
Galton came up with a truly magnificent tea that was 'admirable,
strong and fresh and pure'. This improving vein of brewing
continued through the weekend, and on Sunday morning, he
produced another winning cup of 'very excellent tea (it is true it
had cream)'.

At the end of March the experiments drew to a close, but Galton
must have overlooked something because by November he was
back in the kitchen, brewing up a storm. His notebooks from this
period contain pages of tea-making theory, but the equations and
tables of figures are too fragmentary to make much sense. One
section has the heading, 'To find the capacity for heat of the
teapot'. First, Galton defines his variables:

n = number of ounces of water used

e = excess of its temperature above that of the teapot

t = additional temperature attained by the pot after the water has been poured in

C = required capacity

From this, and with no further explanation, he derives the following equations:

$$C + ne = (C + n)t$$
$$C = n(e - t)/(t - 1)$$

These, and dozens of other opaque explanations, make it impossible to understand his tea-brewing theory. Even Galton seemed to lose his train of thought. After three months of experiments his notes reveal a conspicuous lack of conclusions. But a more coherent summary of his research did turn up in a revised edition of *The Art of Travel* many years later, in which Galton explained all. His ideal cup of tea 'was only produced when the water in the teapot had remained between 180° and 190° Fahr., and had stood for eight minutes on the leaves. It was only necessary for me to add water *once* to the tea, to ensure this temperature.' So keeping the water temperature constant for a measured period of time seemed to be the key to the perfect cuppa. Galton acknowledged that not everyone would share his taste in tea. 'Be that as it may, all people can, I maintain, ensure uniformity of good tea, such as they best like, by attending to the principle of making it, – that is to say, to time, and quantities, and temperature. There is no other mystery in the teapot.'

The tea issue apparently resolved, Galton turned his attention to gold. Two questions had been bothering him for some time. How much gold was there in the world, and would it all fit inside his house? In 1800 the total value of the world's mined gold was estimated at £225 million. Via a protracted series of assumptions and back-of-the-envelope calculations, Galton worked out that all this gold would occupy 3,053 cubic feet. The result came as a shock. Not only would he be able to fit the gold inside his house, he could easily squeeze the whole lot into his dining room, with plenty of space to spare. 'My room,' he declared triumphantly in his notebook, 'without extra window space, but disregarding curve at corners of cornice, would hold more gold than was extant in 1800 by 94 cubic feet'.

Galton's growing appetite for numbers spilled over into all areas of his work. But in print he usually managed to maintain a sense of perspective. The numbers were ever present, but they were never allowed to drown out a practical and beneficial application. 'The Exploration of Arid Countries', a fifteen-page paper published in the newly established *Proceedings of the Royal Geographical Society*, typified the style.

Galton knew from experience all about the problems and frustrations of desert exploration. Water was always the first priority, but the efforts required to find it hampered the overall progress of any expedition. So Galton came up with a novel scheme that would override this element of chance, and provide regular and predictable provisions of food and water along the route. Instead of explorers heading off on their own, they would be assisted by a supporting party whose only job was to set up food caches along the way. This series of camps would provide food and water for the explorers on their return back to base. It was a simple idea that required careful calculations to execute effectively. Characteristically, Galton left nothing to chance, and he gave detailed tables showing the weights and volumes of food and water required based on the distances covered and number of men involved. He envisaged that his new system would benefit those exploring the deserts of Africa and Australia, but it has really come into its own at the opposite end of the climate spectrum. Himalayan exploration, with its extended system of camps and Sherpa support, has wholeheartedly embraced Galton's vision.

In the years that immediately followed his African adventure most of Galton's published work was concerned with travel, exploration, and the finer points of mapping. But his mind was too restless to be confined by a single subject like geography. Besides, it was becoming increasingly apparent that measurement was his true métier, and he was happy to go wherever it took him. So in the late 1850s it was no surprise to find him moving in on meteorology, a subject full of numbers.

In the great days of sail accurate weather forecasting could have saved thousands of lives. It was a fact of which Galton was perhaps more aware than most. In 1853 the *Dalhousie*, the ship that had taken him to South Africa, sank in a violent storm off Beachy Head in the English Channel. All but one of the eighty crew and passengers drowned.

In Britain the job of reporting the weather fell on the shoulders of one man, Captain Robert Fitzroy. An earnest, religious man, Fitzroy had combined a lifelong passion for meteorology with careers in the Royal Navy and the diplomatic service. He was best known to the public as the captain of HMS *Beagle*, the ship that took a young Charles Darwin on his five-year voyage around the world in the early 1830s.

By the mid-1850s Fitzroy was installed as head of the Meteorological Department, a new branch of the Government's Board of Trade. From his office in London he would collect reports from weather stations across the British Isles via the latest telegraph technology. His immediate responsibility was to collate this incoming weather data and draw up simple charts showing wind, air pressure, and temperature.

These maps were necessarily retrospective; by the time they were printed the weather had already moved on. But Fitzroy believed he could go one stage further and actually predict the weather based on the atmospheric data he was receiving. On 1 August 1861 *The Times* began publishing these first, primitive attempts at forecasting.

Fitzroy's forecasts were vague and often inaccurate. This was hardly surprising given the state of meteorological knowledge. The problem was primarily one of scale. While it was known that weather systems evolved over vast distances – thousands of kilometres in many cases – understanding these systems required simultaneous reports from innumerable, widely scattered sources across the globe. In 1860 there was neither the infrastructure nor the organisation necessary to get a glimpse of this bigger picture.

Recognising these shortfalls Galton decided to take matters into his own hands. If there was no integrated system of coordinated weather reporting, then he would organise one for himself, using nothing more high-tech than a paper and pen. In the summer of 1861 he sent off a letter, written in English, French and German, to meteorologists across Europe. The letter was a polite request for a favour. He wanted each meteorologist to supply him with weather data, recorded at three specified times a day, throughout the month of December.

The letter met with a mixed response. While meteorologists in Belgium, Holland, Austria, and Germany seemed only too happy to meet his request, those in France and Italy lacked enthusiasm for the exercise, while the weather men of Sweden and Denmark ignored him completely. As a consequence Galton's European weather data for December 1861 were a little patchy. Nevertheless, he was excited by the general response and eagerly plotted out the wind and barometric pressure readings for each of the December days.

During the first half of the month an area of low atmospheric pressure sat on top of Europe, with the winds rotating about its centre in an anti-clockwise direction. It was a weather pattern familiar to all meteorologists, and one of the few that was understood. The so-called cyclone had first been detected in the tropics, where it could reach a ferocious scale. Tropical cyclones were hurricanes and typhoons in all but name. Up in the higher latitudes of the northern hemisphere it was more common to see the calmer cousins of these revolving storms, but the basic principle was the same: a rotating column of air spiralling inward towards the centre of the depression.

Knowledge of cyclones was key to understanding broad-scale weather patterns. But as Galton was about to discover, cyclones were only half the story. As he continued to plot out his own European weather data, he noticed an abrupt change for the second half of December. The depression had moved away to be replaced by a high-pressure system. And with the change came a complete reversal in the wind direction. The high was surrounded by a rotating column of air spiralling away from its centre. It was something that no one had spotted before. He called it the anti-cyclone.

The name, of course, has stuck. Cyclones and anti-cyclones form a complementary system of air circulation that pretty much defines how our weather works. They are the yin and yang of our atmospheric philosophy. Spiralling, high-pressure air is forced out of an anti-cyclone and 'feeds' the inwardly spiralling, low-pressure air of a cyclone. As the cyclone fills up, air is forced upwards and outwards, replacing the air lost from the anti-cyclone. In a sense, the two pressure systems work together like a pair of inter-connected and alternating bellows, driving air backwards and forwards. We can't sense these subtle differences in pressure, but we can feel their effects in the wind on our faces, and the weather that results.

Not content with one meteorological landmark, Galton sought to illuminate other areas of the weather. Today, when the weather forecast comes on our television screens, we take the weather map for granted: the lines of isobars; the arrows indicating wind speed and direction; the temperature circles; the visual symbols of cloud, rain, and sunshine: they all seem so logical and familiar. But Galton spent years working through a range of pictorial schemes before he came up with one he was happy with. The fruits of his labours appeared in *The Times* on Thursday, 1 April 1875. It was the first public weather map ever to be published, and its style differed little from the maps we see today.

Galton's map may have been clear, concise and a template for the future, but it was still a retrospective one. Forecasting lay beyond even Galton's extensive reach, and he was quick to criticise those who pretended otherwise. By the mid-1860s Fitzroy's

forecasts were beginning to try everyone's patience, generating a
storm of protest from scientists and public alike. Even *The Times*
seemed eager to distance itself from the weather information it was
printing, one reporter commenting wryly that when 'Admiral
Fitzroy telegraphs, something or other is pretty sure to happen'.

Despite his best intentions Fitzroy was struggling to maintain a
public service without adequate knowledge. When the world
ridiculed his naive persistence the criticism took its toll. On
Sunday, 30 April 1865, as he was getting ready for church, Fitzroy
picked up his razor and cut a deep, unforgiving cleft across his
throat.

Fitzroy's suicide was a wake-up call to the weather service. The
Board of Trade appointed a special committee to consider the
future direction of forecasting, with Galton at the helm. A sense of
perspective was needed. Fitzroy's forecasts had been far too ambitious. Any new weather service should have more modest aims.
Storm warnings would continue, but attempts at more general
forecasting would be abandoned for the time being. With Galton in
the hot seat it was no surprise to find the committee also recom-
mending that more weather observations were needed. Automa-
tion was the best way of increasing the volume and speed of
information gathering, and Galton set to work on designing and
building mechanical devices that could record meteorological
information at remote, unsupervised weather stations.

Changes in policy also brought about a change in name. Fitzroy's Meteorological Department became first the Meteorological
Committee and then the Meteorological Council, with Galton an
integral member of both. The Council would later evolve into the
current United Kingdom Meteorological Office. Public weather
forecasts were resumed in 1879 but it wasn't until after the First
World War that people could plan a picnic around them with any
kind of confidence.

Galton's move into meteorology was designed to augment his
interest in geography rather than replace it. Consciously or not, he
was developing his repertoire, building his claim to the title of

Weather man: Galton, Honorary Secretary to the
Royal Geographical Society, 1850s

polymath. But with this diversification of interests there was always the danger that he would spread himself too thinly. The warning signs were already there in the early 1860s, when a string of literary failures revealed cracks in the Galton edifice.

With the success of his earlier books, it was only natural that publishers would come banging on his door, hoping to cash in on his name. When Alexander Macmillan asked him to act as editor and contributor to a new series of travel books, *Vacation Tourists and Notes of Travel*, Galton seemed happy to take up the challenge. The first volume, published in 1860, contained an eclectic mix of contributions. Among articles on pre-revolutionary Naples and the Samaritans of Nablus, Galton wrote an account of his own recent trip to northern Spain, where he had witnessed a total solar eclipse.

While the basic idea behind the book was sound, some of the critics gave it a lukewarm reception, highlighting its ugly lurches of style and sloppy, error-strewn production. *Vacation Tourists* continued the following year, but the second volume fared little better that the first. With reviewers predicting its demise, the series limped on to a third and final volume in 1863.

The following year Galton accepted an offer from publisher John Murray to write a walking guide to Switzerland. The country had become a favourite holiday destination for the Galtons, and after half a dozen holidays there he must have felt competent to write a reliable guide. But an anonymous reviewer in the *Alpine Journal* found the book anything but accurate. Not only was it riddled with grammatical errors, there were mistakes in the times of tours and glaring geographical omissions. These were strange and uncharacteristic blunders for a man noted for his attention to detail. Either Galton had no real interest in the book, or he was struggling to keep up the pace he had set for himself.

Elsewhere, evidence suggested that the pressure of work was really getting to him. Galton's level of involvement with the Royal Geographical Society had risen exponentially since his election to the Council in the mid-1850s. But his promotion to the upper echelons of power had brought him into direct conflict with the Executive Secretary, Norton Shaw. The Council had prospered under Shaw's stewardship, and he was popular with the majority of Fellows. But there were doubts in some quarters about his geographic credentials, and a half-whispered belief that Shaw was an administrator first and a geographer second.

The dispute between Galton and Shaw was ignited by the Society's new publication, *Proceedings of the Royal Geographical Society*, a journal that Galton himself had instigated. While Galton wished to influence the direction of the journal, the editorship fell squarely on Shaw's shoulders. Clearly irritated, Galton spent five years doing his best to wear Shaw down, bombarding him with unreasonable requests and unwanted advice, in the hope that he might relinquish his editorial role. But it was Galton, not Shaw, who began to show the first real signs of fatigue. In April 1862 Galton wrote an angry note to Shaw, chastising him for not reading out a letter at the previous meeting of the Council. In fact the letter had been read, and by none other than Galton himself.

Galton's concerted attack on Shaw backfired badly. Opinion tended to side with Shaw, and while many applauded Galton's ambitious plans for Council reform, his methods won him few friends. Clements Markham, one-time President of the Society, remembered Galton in less than glowing terms. 'His mind was mathematical and statistical with little or no imagination,' he wrote. 'He was essentially a doctrinaire not endowed with much sympathy. He was not adapted to lead or influence men. He could make no allowance for the failings of others and had no tact.' Galton eventually resigned as Honorary Secretary in 1863. Many Fellows argued that his resignation didn't go far enough and wanted to see him removed altogether. But Galton clung on.

The contretemps with Shaw could be put down to a clash of personalities, a classic power struggle in which Galton, the opinionated purist, fought the flawed but popular Shaw. The astonishing memory lapse over the Council letter, however, was a much more worrying sign, and suggested that Galton's mental state was weakening. He was pushing himself too hard, stretching his work load beyond the bounds of safety. Not for the first time in his life his body showed the tell-tale signs of stress and overwork. Heart palpitations, dizziness, and nausea had followed him on his upward climb through the professional ranks.

His resignation as Honorary Secretary left a small hole in his schedule. It was a chance to step back from the gruelling regime he had set himself. But time off had become an alien concept to Galton. If a gap appeared in his diary he was compelled to fill it immediately. So while Society business took a back seat, Galton found himself other editorial duties to attend to.

It is a measure of how far Galton had come as a scientist that he was able to slip comfortably into a new literary venture that combined some of the finest intellects of the day. In 1864 biologist Thomas Henry Huxley, philosopher Herbert Spencer, physicist John Tyndall, astronomer Norman Lockyer, historian John Seeley, and author Charles Kingsley joined Galton and many other distinguished names on the editorial board of *The Reader*, a new weekly magazine of science, literature and art. Science, in particular, was poorly catered for in the popular press and *The Reader* set out to fill this niche, communicating new scientific developments and ideas between the educated classes.

Blessed with such a heavyweight intellectual presence, and the tacit support of many more great scientific names, the magazine looked set for a long and distinguished life. But it was beset with problems from the start. It was one thing to have the backing of the best Victorian minds, quite another to get them to produce intelligible articles on time. With its mostly voluntary editorial staff, always busy on other things, the magazine turned into a shambolic, rudderless affair. Stolid and dull, it finally self-destructed in 1866 amid a blaze of indifference. Three years later Norman Lockyer reinvigorated it under a slightly different guise. *Nature*, embracing a purely scientific format, went on to much greater things, and now regards itself as *the* international journal of science.

With Galton consumed by work, Louisa was seeing very little of her husband. Each year she made a brief summary of events in her 'Annual Record', and her entry for 1860 hints at her frustrations. 'Went to Leamington and Claverdon early in January. Let our house from May 1st for two months. Took lodgings at Richmond. Constant bad weather. Home sick to a painful degree and Frank almost always away in London or at Kew Observatory.'

Only on their regular holidays could she guarantee any time together. But even then, her husband's thoughts were often on other matters. In 1860, for instance, the couple travelled to Luchon in the French Pyrenees. A thousand miles from the societies and committee meetings of London, amid the dramatic mountain scenery, it seemed like the perfect place to relax and reaffirm their marital bond. But Galton had other ideas. Having checked into their hotel, he slipped away into the night, leaving Louisa alone once more.

It was in the Pyrenees that Galton became 'bitten with the mania for mountain climbing'. This particular night, however, he had other things on his mind. He had heard about a new sheepskin sleeping bag that French soldiers had just started using in their smuggling patrols of the Pyrenean mountain passes. Keen to try one out for himself, he left Louisa behind and went halfway up a mountain to give it a test drive.

> A heavy storm was gathering, but before the evening closed and before the storm broke, I had time to find a good place on a hill some 1,000 feet or more above Luchon, and there to await it inside my bag. Nothing could have been more theatrically grand. The thunder-clouds and the vivid lightning were just above me, accompanied by deluges of rain. Then they descended to my level, and the lightning crackled and crashed about, then all the turmoil sank below, leaving a starlit sky above.

Although Galton's relationship with Louisa seemed distant, it was a gulf that could have been bridged by a child, a new being in whom they could genuinely share their lives. But a decade into their marriage there were still no signs of a baby, and with Louisa now into her forties, time was running out. The sterility of the marriage must have weighed heavily on Galton's conscience. Although infertility was something that ran in their families – Galton's brothers and Louisa's sisters were all childless – there were other, more sinister possibilities. Galton couldn't be sure, but his youthful indiscretions in Beirut, and the medical problems that resulted, could have contributed to an infertility all of his own making.

By 1865 he seemed to have resigned himself to a childless life. 'It seems natural to believe', he wrote, 'that a person who . . . does not happen to have children, should feel himself more vacant to the attractions of a public or literary career than if he had the domestic cares and interests of a family to attend to.' It was a rare display of candour from a shy and intensely private individual – the public justification of a man tortured by the inability to have a son or daughter of his own. For Louisa, the future must have looked bleaker than ever.

8

Extreme States

The religious instructor in every creed is one who makes it his profession to saturate his pupils with prejudice.

Francis Galton

In the summer of 1860 Oxford played host to the annual meeting of the British Association for the Advancement of Science. The conference had been eagerly anticipated by all those who enjoyed a good argument. Darwin's *The Origin of Species* had appeared six months earlier, and the controversy surrounding its publication was in full swing. Darwin himself had never set out to be deliberately provocative. But in his quiet and unassuming way, he had rubbished the theological view of man as the pinnacle of creation, relegating him to a mere bit-part player on the earth's stage. For all our efforts to distance ourselves from the animals, he argued, we were just another twig in the tree of life, sitting on a branch with the rest of the apes. It was sensational stuff and caused uproar across the country. The Oxford conference was an opportunity to let off some steam, and a chance for a new breed of Darwinian converts to confront the old religious order.

In the event, the meeting more than lived up to its billing, and has gone down in history as the seminal Darwinian debate. On one side, representing the case for God, stood the Bishop of Oxford, Samuel Wilberforce; on the other, representing the case for science, was 'Darwin's Bulldog', Thomas Henry Huxley. Egged on by the Christian anatomist and palaeontologist, Richard Owen, it was Wilberforce who threw the first punch, mocking Darwin's theory and taunting Huxley with the question of whether 'it was through his grandfather or his grandmother that [he] claimed his descent from a monkey'. Huxley, unruffled, came back at Wilberforce with a devastating riposte: 'If . . . the question is put to me, would I rather have a miserable ape for a grandfather or a man highly endowed by nature and possessed of great means of influence and yet who employs these faculties and that influence for the mere purpose of introducing ridicule into a grave scientific discussion, I unhesitatingly affirm my preference for the ape.'

Galton was somewhere among the crowd on that famous day, although, unaccountably, his autobiography contains no specific mention of the event. Nevertheless, there was no doubt where his sympathies lay. Darwin's ideas had triggered a revolution in his thinking. *The Origin of Species*, he claimed, 'made a marked epoch in my own mental development, as it did in that of human thought generally. Its effect was to demolish a multitude of dogmatic barriers by a single stroke, and to arouse a spirit of rebellion against all ancient authorities whose positive and unauthenticated statements were contradicted by modern science.'

To the dismay of the deeply devout Louisa, the lingering traces of Galton's religious feelings disappeared into dust. He felt possessed by a new and profound sense of clarity. In a heartfelt letter to his half-cousin he was ecstatic: 'Pray let me add a word of congratulation on the completion of your wonderful volume . . . I have laid it down in the full enjoyment of a feeling . . . of having been initiated into an entirely new province of knowledge which, nevertheless, connects itself with other things in a thousand ways.'

The 'survival of the fittest' had a resonance that went beyond discussions of the evolutionary history of life. In an industrial, post-Enlightenment world, it was a powerful and persuasive idea that could and would be used to justify all manner of social, cultural, and political manifestos, often to the embarrassment of

Darwin himself. Extreme *laissez-faire* capitalism was the stereo-typical interpretation of the Darwinian metaphor, the state pas-sively looking on as individuals and businesses fought among themselves for economic supremacy. But social darwinism was a loose, umbrella term that covered a variety of political persuasions. Many liberals, for example, would argue for the abolition of class barriers in the name of social darwinism. Only then, they insisted, would there be a level playing field from which a true biological meritocracy could emerge.

The Origin of Species may have been the catalyst behind a new brand of social thinking, but the boundaries between science and society had begun to blur long before Darwin burst onto the scene. Darwin himself was a product of his age as much as a pioneer, and some of his book's fundamental themes borrowed heavily from the caravan of ideas stamping its way through the Victorian era. The evolutionary philosophy of Lamarck, Malthus's gloomy predic-tions on overpopulation, and Herbert Spencer's 'struggle for existence' were already embedded in the Victorian consciousness.

But what set Darwin apart from his contemporaries – and made *The Origin of Species* such an important book – was the scientific authority he brought to his subject. Drawing on years of painstak-ing research, Darwin had come up with a compelling mechanism to explain how evolution occurred. His theory of evolution by natural selection made no appeal to God or metaphysics, and was as straightforward as it was brilliant. Animals and plants, he argued, produce more offspring than their environment can sup-port. This inevitably leads to a struggle between individuals for food, mates, and living space. Heritable differences between in-dividuals means that some will stand a better chance of survival than others. Each generation, natural selection sifts the winners from the losers, eliminating the unfit and adapting individuals to their environment.

Despite its apparent simplicity, Darwin's argument caused con-siderable confusion. Looking back from the pages of his auto-biography, Galton doubted whether 'any instance has occurred in which the perversity of the educated classes in misunderstanding what they attempted to discuss was more painfully conspicuous'. The meaning of the simple phrase 'Natural Selection' was he added, 'distorted in curiously ingenious ways, and Darwinism was attacked, both in the press and pulpit, by persons who were manifestly ignorant of what they talked about'. In contrast to the muddled masses, Galton claimed he had no problem connecting with the central message contained within *The Origin of Species*. He 'devoured its contents and assimilated them as fast as they were devoured'.

Like all fresh converts, Galton was zealous in the pursuit of his new craze. Publicly, he seemed to be consolidating his position as a geographer and meteorologist of world renown. But behind this public face he was working himself into a frenzy, taking the bits he liked best from Darwin and moulding them into radical new shapes and forms. Darwin had used the domestication of animals and plants to illustrate the way in which evolution by natural selection produces changes in the form and features of species. People had become agents of selection, adapting animals and plants to their own particular needs. By controlling which individuals were allowed to breed over many generations, desirable characteristics could be exaggerated and unwanted features suppressed.

Intrigued by the apparent plasticity of life, Galton dared to imagine an audacious breeding experiment of his own. This time, however, the subjects for selection would not be cattle, pigs, or maize, but people:

> If a twentieth part of the cost and pains were spent in measures for the improvement of the human race that is spent on the improvement of the breed of horses and cattle, what a galaxy of genius might we not create! We might introduce prophets and high priests of civilisation into the world, as surely as we can propagate idiots by mating *crétins*. Men and women of the present day are, to those we might

hope to bring into existence, what the pariah dogs of the streets of an Eastern town are to our own highly-bred varieties.

Galton gave free rein to his fantasy. He imagined a Utopian state in which only the most able in mind and body would be allowed to procreate. Each year the mental and physical abilities of the country's young people would be assessed. This annual exam would be a distillation exercise, a test to find the best men and women in the land. Once selected, these individuals would be encouraged to mingle, to marry, and, especially, to mate. In the event of any motivational difficulties between couples, lavish weddings and generous handouts would be provided by the state as additional incentives. These chosen people would become the seeds from which a new race of supermen and superwomen would grow.

As a born-again Darwinian, Galton had dreamed up a social experiment in human domestication. If farmers could use artificial selection to exaggerate desirable features in animals and plants, then why could society not do the same to the minds and bodies of men? He would not coin the term eugenics until 1883. But in 1864 a rough outline of his new hereditary philosophy was already in print.

'Hereditary Talent and Character' appeared in *MacMillan's Magazine* as a two-part article in November 1864 and April 1865. In its twenty or so pages Galton wove together disparate ideas about heredity that came together in his vision of a eugenic paradise. In one sense there was nothing very original in Galton's claim that heredity was the source of all human differences. After all, heredity was an acknowledged fact in plants and animals. Heritable variation between individuals was the fuel of evolution, the diversity from which natural selection chose its winners and losers. Without it, animal and plant breeders wouldn't have got very far in their attempts at domestication.

But Galton's assertion went beyond a discussion of purely physical attributes into far more controversial realms. He argued

Reborn: Galton, aged forty-two, shortly before the publication of
'Hereditary Talent and Character'

that if physical features were under some kind of hereditary
control, then so too, in all likelihood, were mental characteristics.
The great spectrum of human aptitudes, outlooks, and abilities, he
insisted, was also an inborn element of our being.

The origins of human nature had kept philosophers in business
since the dawn of civilisation. For many, mere talk of the human
mind introduced a metaphysical aspect to the equation, rendering
the whole subject off limits to scientific inquiry. Outside the
theological debates, that trio of philosophical heavyweights John
Locke, David Hume, and John Stuart Mill championed the view
that the human mind was a 'blank slate', inscribed by experience.
In 1865 few intellectuals wholly endorsed this strictly environ-
mentalist viewpoint. But in the absence of any scientific insight,
the subject was a bit of a free-for-all, a rich oasis of opinion amid a
featureless desert of fact.

Galton, however, declared he had made a breakthrough, and he
was impassioned, almost fanatical in his assertions. He claimed to

have been alerted to the idea of hereditary talent while he was still at university: 'I had been immensely impressed by many obvious cases of heredity among the Cambridge men.' Many of the top men of his year – the so-called senior wranglers and senior classics – had close relatives who had also been top of their respective years. It seemed highly unlikely, Galton argued, that these family clusters of eminence were due to pure chance. The more likely explanation, he asserted, was that eminence was strongly hereditary.

In 'Hereditary Talent and Character' Galton tried to turn a hunch into hard facts. Beyond the hyperbole he had actually come up with an intriguing scientific approach to the problem. His experimental 'proof' was based on statistical considerations of eminent men and their relatives. Referring to Sir Thomas Phillips's *The Million of Facts*, a popular reference work of the day, Galton had gathered together a list of 330 eminent men of science and literature. When he scrutinised the family pedigrees of these men he found that one in twelve had a relative who was also on the same list. From his calculations, he deduced that a person was almost 300 times more likely to be eminent if they had a relative who was also eminent. To Galton, the evidence of hereditary influence was overwhelming.

Galton adopted a hectoring, rhetorical tone throughout the article, almost as if he had made his mind up on the issue of hereditary influence *a priori*. But while he did provide a convincing case that ability had a tendency to run in families, the question of whether this was caused by a shared heredity or a shared environment remained unresolved. If eminence was a consequence of wealth, social status, and parental influence, then it might produce similar patterns in the pedigrees. Galton, however, didn't see things that way, and his efforts to gloss over the environmental issue revealed more about his own prejudices than any solid scientific reasoning.

The hereditary conclusion was convenient for Galton. Eugenics – the next phase of his argument – was entirely dependent upon it. Without inborn variation any attempt at selective breeding would be a fruitless exercise. Of course, the alternative reality – one in which environmental influences shaped human differences – would not necessarily undermine the ideal of human improvement, but it would imply an altogether different kind of Utopia from the one to which Galton aspired. If ability was purely an acquired character, then human improvement would depend on social and educational reforms rather than selective breeding.

Galton made the philosophical leap from hereditary influence to eugenics as if it was the most natural thing in the world. But considering the drastic nature of his proposals, he seemed a little reticent when it came to explaining why such measures were needed in the first place. What was so bad about the human race that it required such serious hereditary tinkering? Perhaps progress – the mantra of the Victorian age – was reason enough. But Galton's explanation, when it eventually came, revealed far more specific concerns: 'The average culture of mankind is become so much higher than it was, and the branches of knowledge and history so various and extended, that few are capable even of comprehending the exigencies of our modern civilisation; much less of fulfilling them.' Few would have denied that society was becoming more complex. But Galton seemed to mutate this simple fact into something much more melodramatic. He insisted – without deigning to provide any evidence – that human mental capacities were failing to keep up with the cultural transformations taking place around them. If our biological evolution was falling behind our cultural evolution, he screamed, then surely it was irresponsible not to intervene?

Galton's rationale for a eugenic future seemed to be rooted in a very old-fashioned view of the intellectual. Galton had grown up in a world of polymaths. Many of the scientists of his generation, men like Sir John Herschel, and his old Trinity College master, William Whewell, were never constrained by one subject alone. But as the nineteenth century grew older the omniscient mind was yielding to the escalating weight of knowledge. If the polymath was an endangered species then eugenics was Galton's strategy of conservation.

Galton didn't go into too much detail regarding the practicalities of his plan. That would all come later. But he did sketch a rough outline to illustrate how such a policy might take effect. He considered two groups of people, A and B, of which 'A was selected for natural gifts, and B was the refuse'. Marriage would only be allowed between individuals from the same group. The chosen ones in group A would be encouraged to marry early, have children, and lots of them, while the reproductive ambitions of the rejects in group B would be 'discouraged' or 'retarded'. After many generations, these differential rates of reproduction would lead to the complete extinction of group B.

While Galton's aim was clear – the improvement of mental ability through selective breeding – he obviously anticipated concerns about the unwanted side effects that such a policy might cause. There was, it seems, a popular belief at the time that what eminent men had in intellect, they lacked in sexual prowess and physical strength. If there was any truth to the claim then selecting for ability might lead to the creation of a race of brainy, asexual weaklings. But Galton dispelled the myth with some delicately handled, if sweeping revelations: 'I, however, find that very great men are certainly not averse to the other sex, for some such have been noted for their illicit intercourses, and, I believe, for a corresponding amount of illegitimate issue.' On the subject of strength he was much more effusive: 'Men of remarkable eminence are almost always men of vast powers of work . . . Most notabilities have been great eaters and excellent digesters, on literally the same principle that the furnace which can raise more steam than is usual for one of its size must burn more freely and well than is common.'

For the remainder of the article Galton brought together a ragbag collection of half-developed, if intriguing ideas, all of which radiated from his hereditary theme. First on the list was religion. The Church, he fumed, had a lot to answer for. Its medieval policy of appropriating intelligent, talented men for a life of celibacy was nothing less than a crime against the very principles of eugenics.

Generations of abstinence from able minds must, he insisted, have been a continual drain on the hereditary stature of humanity. Here, Galton was evidently building on his eugenic justification. It was bad enough that our biological evolution was lagging behind our cultural advance. But having already binned a huge vat of hereditary talent, religion had only made things worse. A heavy dose of eugenic medicine was obviously the only way to get our biology back on track.

Galton must have been feeling a great sense of release and retribution after his own religious emancipation because he delivered another broadside a few rambling pages later. Spiritual sentiments, he argued, were nothing more than a relic of our evolutionary ancestry: 'All evidence tends to show that man is directed to the contemplation and love of God by instincts that he shares with the whole animal world, and that primarily appeal to the love of his neighbour.' Here, Darwin's influence was at its most transparent. Galton had quickly acquired the habit of putting humans and the rest of the animal kingdom on the same evolutionary stage: 'We are still barbarians in our nature, and we show it in a thousand ways.' Civilisation was a thin veneer that barely concealed the cracks of our animal instincts. Our sense of original sin, he believed, sprang from the conflict between these two opposing forces.

With the God issue taken care of Galton turned his attention towards racial differences. Here again, he declared, the evidence for hereditary influence was overwhelming. Rather than dwell on the obvious physical differences between people, Galton preferred to concentrate on supposed differences in racial character. According to unnamed, yet 'excellent observers', the North American Indians were 'naturally cold, melancholic, patient and taciturn . . . They nourish a sullen reserve, and show little sympathy with each other, even when in great distress.' Contrast this, Galton urged, with the character of the West African. 'The Negro has strong impulsive passions, and neither patience, reticence nor dignity. He is warm-hearted, loving towards his master's children, and idolised by the children in return. He is eminently gregarious, for he is always jabbering, quarrelling, tom-tom-ing, or dancing.'

That such character differences were rooted in heredity was evident, Galton explained, when you looked at the American continent. American Indians occupied all kinds of environments, from the polar regions of Alaska and Canada in the north, through the temperate zones of the United States, the tropics of Central and South America, all the way down to the cold tundra of Tierra del Fuego in the south. Throughout the continent populations had been exposed to a wide variety of political and religious systems. Almost every country in Europe had left its own indelible mark in some way or another. And yet despite this diversity of environmental influences, Galton insisted, the character of the American Indian was essentially uniform.

After his character assassination of the American Indians Galton moved swiftly on to a slur of its more recent residents. It was a certain kind of person, he said, that had wanted to emigrate to America. The white settlers were obviously a varied bunch, but there was one aspect of their nature that they had all shared: 'Every head of an emigrant family brought with him a restless character, and a spirit apt to rebel.' It was no surprise, therefore, to find that the character of a country germinated by these kind of seeds had turned out the way it had. According to Galton's assessment, Americans were 'enterprising, defiant, and touchy; impatient of authority; furious politicians; very tolerant of fraud and violence; possessing much high and generous spirit, and some true religious feeling, but strongly addicted to cant'. Galton, of course, never had and never would set foot on American soil.

Given the number of hornets' nests he trampled over so un-flinchingly, it is surprising to discover that 'Hereditary Talent and Character' aroused barely a response. Perhaps the tone of the article was too off-putting. It was certainly an audacious, ranting discourse, and always threatened to teeter into fanaticism. Galton wrote with an unnatural fearlessness that sometimes took him beyond arrogance into the distorted world of the fundamentalist.

There was a sense of desperation in his tendency to overstate his cause.

In truth, his mental state was in a perilous condition. Ever since his return from Africa in 1852 he had maintained an increasingly hectic schedule. 'During the whole of this interval,' he recalled in his autobiography, 'I frequently suffered from giddiness and other maladies prejudicial to mental effort.' His condition would usually improve whenever he took a holiday abroad, or plenty of outdoor exercise, but Galton didn't really seem to digest this simple equation. For years work had threatened to overwhelm him. Relatives urged restraint, but he seemed either unwilling or unable to heed their advice. He knew where all this could lead. His experience at Cambridge had taught him that. But somehow he seemed powerless to change his course.

9

On the Origin of Specious

I often feel that the tableland of sanity upon which most of us dwell, is small in area, with unfenced precipices on every side, over any one of which we may fall.

Francis Galton

Galton was alone in the bath, thinking about a letter from his mother. 'Do, my love, be persuaded to lay aside all ambition and give yourself a complete year's rest and perhaps it may be the saving of your life. Come to me whenever you like and you shall have perfect quiet. My mind is with you by day and by night. I long to have you with me.'

Moving his hands up over his face, he ran his fingers through an imaginary fringe. The surface of his scalp was as bare as the Namibian desert. Ever since his teens, follicles had been disappearing like trees in a forest fire. His brain had become a furnace, fuelled by facts. The heat it produced was obviously incompatible with a full head of hair. That was why women always had so much hair. Less fire in the furnace meant more hair on the head.

Galton grabbed his copy of *The Times* and slid below the surface, taking the paper with him. Beneath the rippling bath water he devoured the latest dispatches from Fleet Street. It was bleary at first, but with his home-made pair of submariner's spectacles he could just about bring the words into focus. There was an article about Abraham Lincoln's funeral, but he was easily distracted. Something else was calling for his attention. A great sickness was washing over him. He could hear the echo of his mother's voice, but the words were drowned out by the thump of an aching heart and a brain still urging reason. At last he lifted his head up and out of the water to snatch greedy gulps of steaming air.

The house was silent. All he could hear was the deafening sound of his own heavy breathing. He pulled his limbs into his body, cowering in his bath. Where was Louisa? He thought he heard a baby crying. But he was alone. Childless and alone.

Page four of *The Times* contained an intriguing article on William Booth and the Salvation Army. It looked like perfect underwater reading, so he went diving again. But when he was halfway down the page the middle three columns reached super-saturation, detached themselves from the rest of the paper, and floated off in the direction of his feet. By the time he'd retrieved them the sickness was on him again.

Back above the surface he rested his chin on his chest, and then shifted it a couple of centimetres to the left until he thought he'd found the perfect line of symmetry. But the view depressed him. The svelte African explorer had grown fat on endless rounds of committee meetings and luncheons at the Athenaeum.

Through the open window of the bathroom Galton caught sight of a hippo coming up the street towards him. The animal looked familiar. It was moving fast, its belly swaying like an unruly piece of offal. With its mouth wide open he could just make out some words tattooed in black ink on its massive incisors. They seemed to be written in his own hand. He reached for his gun but there was no gun, so he recoiled further into the recesses of the tub. He could only watch in horror as the postman turned the corner of Rutland Gate and met the hippo head on. But the postman was well armed. With one vicious swipe of his postbag he dispatched the hippo into the crown of a horse-chestnut tree in nearby Hyde Park.

In February 1866, a few days short of his forty-fourth birthday, Galton found himself in a delicate mental state. This was a much more serious breakdown than the one that had left him catatonic at Cambridge:

> The warning I received in 1866 was more emphatic and alarming than previously, and made a revision of my mode of life a matter of primary importance. Those who have not suffered from mental breakdown can hardly realise the incapacity it causes, or, when the worst is past, the closeness of analogy between a sprained brain and a sprained joint. In both cases, after recovery seems to others to be complete, there remains for a long time an impossibility of performing certain minor actions without pain and serious mischief, mental in the one and bodily in the other. This was a frequent experience with me respecting small problems, which successively obsessed me day and night, as I tried in vain to think them out. These affected mere twigs, so to speak, rather than large boughs of the mental processes, but for all that most painfully.

His recovery was long and painful, and not helped by his attempts to rush himself back to health. In early 1866 he and Louisa took an extended holiday to the Italian lakes. When they returned in June, Galton already had half an eye on the forthcoming meeting of the British Association in Nottingham. He had agreed to give a talk at the meeting and, ill health or no, he felt obliged to go. Only when he got to Nottingham did he realise how much he'd understimated his mental frailty. Overwhelmed by the hustle and bustle of the conference, he hastily arranged for someone else to read his paper and quietly slipped away.

Acting on the advice of his mother and sisters, Galton withdrew from public life for the next three years. The hectic London schedule was replaced by a more sedate itinerary. Both he and Louisa knew the route back to mental harmony: it went from

England to Italy, via Switzerland, Austria and Germany. Travel and physical exercise went a long way towards ameliorating the worst of his symptoms. But the illness proved difficult to shake off. There were numerous occasions throughout 1867 and 1868 when he seemed to be clawing his way out of the darkness, only to slip and fall back into the abyss.

Galton had prescribed himself an extended continental tour, but this was still not a holiday in any objective sense. Wisely or not, he used his time out to catch up on some reading. This rarely included novels, for he frequently fought shy of fiction. But it did include notes, ideas, and correspondence relating to his favourite theme. Having given up on having a baby of his own, heredity was now officially his adopted child.

Three years of convalescence did little to soften Galton's stance on heredity. As the title of his 1869 book *Hereditary Genius* suggested, he was carrying on very much from where he had left off. The book was an attempt to clarify and consolidate the ideas he had raised in 'Hereditary Talent and Character', reiterating his central thesis that men were great because they were born that way, and not because of their upbringing. 'I propose to show in this book', he began, 'that a man's natural abilities are derived by inheritance, under exactly the same limitations as are the form and physical features of the whole organic world. Consequently . . . it would be quite practicable to produce a highly-gifted race of men by judicious marriages during several consecutive generations.'

To prove his point he resorted to the same methodology he'd used four years earlier. Family histories were again the mainstay of his thesis. This time, however, he obviously felt that he needed to bring a lot more to the table to make a convincing case. The vast bulk of the book was taken up with the detailed family pedigrees of eminent men. The men themselves were divided up according to their professions. Appropriately enough, Galton opened up his argument for the prosecution by considering the judges of England.

Perhaps it wasn't too surprising to find that these stalwarts of society were rich in eminent relatives. For example, Sir Christopher Milton, Justice of the Common Pleas under James II, had a slightly more famous brother, John Milton, the poet and author of *Paradise Lost*. And Sir Henry Gould, Justice of the Queen's Bench under Queen Anne, had two eminent grandsons, Henry Fielding, the novelist, and another Sir Henry Gould, Justice of the Common Pleas under George III. Pages of ancestral information revealed similar patterns. Galton's conclusion, telegraphed from the outset, was that eminent judges had relatives who were also eminent far more often that you would expect by chance.

While many of the judges had eminent mothers, daughters, or brothers, Galton discovered that eminent uncles, aunts, and nephews were comparatively rare. In other words, the ties of eminence seemed to wither as the relationship became more remote. This was an observation that had not come out of his previous study, and was something which, Galton insisted, merely reinforced his hereditary explanation.

Having perfected his formula on judges, he repeated it for other professions. Statesmen, commanders, writers, scientists, poets – 'a sensuous, erotic race, exceedingly irregular in their way of life' – musicians, painters, divines, and the senior classics prize winners of Cambridge all got the Galton treatment. He even extended his umbrella of eminence to include outstanding physical ability, topping off his trawl through the archives with short and slightly peculiar chapters on eminent oarsmen and the wrestlers of northern England.

By the end of the book, the reader would have known a little more than they needed about the great-grand-nephews and second cousins of some of the most famous and not so famous names in history. The book's format made it a bit of a chore to wade through, but for those who persevered, there were occasional rewards. For instance, William Harvey, the physician who discovered the circulation of the blood, was described by Galton as 'a little man with a round face, olive complexion, and small black eyes full of spirit. He became gouty, and acquired fanciful habits. He lay in bed thinking overmuch at night time, and slept ill . . . His relationships show sterling ability.'

There's no doubt that Galton turned up some astonishing findings. The Kennedys, for example, were a remarkable bunch who had made the senior classics prize at Cambridge their own. Benjamin Kennedy was the first member of the family to win it in 1827, followed shortly afterwards by his two brothers, Charles and George, in 1831 and 1834 respectively. Then there was a lull, but the family returned to winning ways when Benjamin's nephew gained the prize in 1868. The Butlers were not far behind the Kennedy family in the eminence stakes. But when you realise that this was the same Butler family into which Galton had married, it rather diminishes the impact of their achievements.

Galton named many of his distant in-laws, but he found no place for his wife, Louisa Butler, a woman of no small intellect. Women were almost completely absent from the book. Obviously much of this had to do with the fact that women had long been denied the social and professional opportunities afforded to men. But Galton added his own glass ceiling by arguing that 'there exists no criterion for a just comparison of the natural ability of the different sexes'. Unsure of how to deal with women, he felt that the best solution was to ignore them.

Despite its extended length the book suffered from the same fundamental weaknesses as had afflicted 'Hereditary Talent and Character'. The first problem was what was meant by the term 'eminence'. Since the word had no accepted scientific meaning, it was impossible to apply objectively. Sometimes Galton resorted to biographical dictionaries to find his lists of eminent men, at other times he chose the men himself. But either way, his methods exposed him to justifiable accusations of prejudice. In the chapter on judges, for example, he referred to Sir John Singleton Copley, a Lord Chancellor under Queen Victoria. Galton added Copley's father to the list of eminent relatives, but his justification seemed a little weak: 'A painter, and an eminent one, judging from the prices that his pictures now fetch.'

Galton's attempts at objectivity were not helped by his deep and transparent admiration for establishment figures. When it came to judges, for instance, he offered more than a hint that he might be prejudging his legal men: 'They are vigorous, shrewd, practical, helpful men; glorying in the rough and tumble of public life, tough in constitution and strong in digestion, valuing what money brings, aiming at position and influence, and desiring to found families.' Sometimes, his passion for able men seemed in danger of boiling over: 'A collection of living magnates in various branches of intellectual achievement is always a feast to my eyes; being, as they are, such massive, vigorous, capable-looking animals.'

While many of the names on Galton's lists were indisputable – few people, for instance, would question his inclusion of Mozart, Bach, and Beethoven – others seriously stretched the credulity of his audience. In the chapter on scientists, for example, Galton found no place for such household names as Faraday, Dalton, or Joseph Priestley. Even Leonardo da Vinci couldn't get onto Galton's list of eminent men. On their own, these omissions meant nothing. But viewed in the context of what Galton did include, things started to look extremely suspicious indeed. Buffon, the French naturalist, did get a mention. Nothing controversial in that. But then Galton unaccountably added Buffon's son to his list of eminent men. Here is a man whose 'abilities were considerable, and his attachment to his father was extreme. He was guillotined as an aristocrat.' Galton admitted in the introduction to the book that he was deeply conscious of the imperfection of his work, 'but my sins are those of omission, not of commission'. Sometimes, the evidence suggested otherwise.

To be fair, Galton recognised the inherent difficulties in defining eminence and looked elsewhere for more objective measures of ability. In his search he had stumbled across the work of Adolph Quetelet, a Belgian astronomer. Quetelet had become an enthusiast of a new and curious statistical phenomenon known as the

'law of deviation from an average'. That Quetelet was playing up
any statistical theory was, in itself, a startling fact. In the mid-
nineteenth century statistical laws were about as common as
dodos. In fact statistics was hardly a subject at all. The word
statistics literally meant 'state numbers', and was normally used
with reference to trade figures, population sizes and manufacturing output. In the absence of any theoretical framework, people
used common sense to help them interpret these numbers. This
was not necessarily a bad thing. At its simplest, modern theoretical
statistics is little more than a mathematical endorsement of common sense. But as Quetelet and Galton both discovered, statistics
can also take you places that common sense can't go.

Quetelet had made his name as Belgium's Astronomer Royal,
but it was a more general passion for measurement that brought
him into the fledgling field of theoretical statistics. In fact it was
with people, not planets or stars, that he demonstrated the law of
deviation from an average. Quetelet had gathered information on
the chest sizes of over 5,000 Scottish soldiers. Having obtained his
measurements he organised them into a table showing the frequency of each chest size. Four men, for instance, had the smallest
chest size of thirty-three inches, thirty-one men had a chest size
of thirty-four inches, 141 men had a thirty-five-inch chest, and
so on, up to the maximum chest size of forty-eight inches. When
Quetelet made a graphical representation of this frequency data he
found that the distribution took on a striking visual form. The data
points traced out a symmetrical bell-shaped curve. In other words,
the chest sizes of most men clustered around the average – about
forty inches – with numbers decreasing rapidly above and below
this size.

If this new law had only been useful in its description of the chest
sizes of Scottish men it probably wouldn't have had much of a
future. But what was so startling about the law was its ubiquity.
Quetelet could have measured the brain weights of politicians, the
diameters of octopus eyes, or the penis lengths of Mediterranean
rabbit fleas to illustrate his point. In each case he would have

uncovered the same bell-shaped curve. Little wonder that this ubiquitous law became known as the 'normal distribution'.

Impressed by the generality of the new law, Galton wondered whether mental ability might be similarly distributed. If it was true for human heights, weights, and all manner of other measurements, Galton saw no reason why intelligence should not follow the same bell-shaped curve. In 1869 there was no standardised measure of human intelligence; the IQ test was still decades away. So Galton went looking for surrogate measures of intelligence, and came up with the results of the 1868 admission exam to the Royal Military Academy at Sandhurst.

Using his newly acquired statistical theory, Galton could predict what the distribution of marks would look like if they had conformed to a perfect normal distribution. In other words, he could test the coincidence of his own observations – the actual exam marks – with this idealised expectation. Nowadays, statisticians can use mathematics to test the strength of the coincidence but, in 1869, eyeballing the data and making an educated guess was about all you could do.

Galton was convinced that his data lived up to expectations. Exam results at least approximated a normal distribution: most of the candidates achieved average marks, while an extreme few achieved very high or very low marks. 'There is, therefore, little room for doubt, if everybody in England had to work up some subject and then to pass before examiners . . . that their marks would be found to range, according to the law of deviation from an average, just as rigorously as the heights of French conscripts, or the circumferences of the chests of Scotch soldiers.' If examination marks were normally distributed then Galton felt sure that intelligence would follow the same general pattern.

Armed with his assumption of normality, Galton proceeded to divide the entire male population of England – all fifteen million of them – into fourteen classes of intelligence. His scale included seven classes above the average, ranging from class *A* up to the highest class, *G*, and seven classes below the average, running from class *a* down to the lowest class, *g*. Galton provided his own idiosyncratic analysis of how his purely theoretical distribution might translate into real life. Most people – more than four-fifths – were in 'the four mediocre classes, *a, b, A, B'* around the average. By mediocre, Galton was referring to 'the standard of intellectual power found in most provincial gatherings'. Moving up the scale, class *C* 'possesses abilities a trifle higher than those commonly possessed by the foreman of an ordinary jury', while class *D* contained the 'mass of men who obtain the ordinary prizes of life'. Galton obviously couldn't think up a social analogy for the *E* class because he moved gracefully into the right-hand tail of the bell curve, into the rare and distinguished realm of classes *F* and *G*. These superior beings were the meat and drink of *Hereditary Genius*, the elite who would form the breeding stock of any future programme of eugenics.

Having explored the highs, Galton plunged himself into the lower end of his intelligence scale. By the time he had reached class *f* he was 'already among the idiots and imbeciles'. He had a lot to say about idiots, and saw a clear distinction between the accidental idiot and the natural-born idiot. Just in case the reader needed a more graphic illustration of his message, he served up another one of his characteristic analogies: 'I presume the class *F* of dogs . . . is nearly commensurate with the class *f* of the human race, in respect to memory and powers of reason. Certainly the class *G* of such animals is far superior to the *g* of humankind.'

Beyond social stereotyping, Galton's bell curve also gave his racism a convenient scientific expression. In one of the more disturbing chapters in *Hereditary Genius*, entitled 'The comparative worth of different races', Galton decided it would be an interesting exercise to rate various races on his scale of human intelligence. Blacks came out badly. The average intellectual standard of the African Negro was, he argued, about two grades below the Anglo-Saxon. But when you looked for the evidence behind this assessment there was none to be found. In its place there was

Galton's four-point plan to blind prejudice. Point one: he couldn't think of any famous, eminent blacks. Point two: a vapid rephrasing of point one. Point three: 'It is seldom that we hear of a white traveller meeting with a black chief whom he feels to be the better man.' And finally, point four: 'The number among the negroes of those whom we should call half-witted men is very large . . . I was myself much impressed by this fact during my travels in Africa.' So there we have it. Rigorous, exhaustive, scientific proof. It was quintessential Galton fare. Throughout *Hereditary Genius* he perfected an intriguing didactic style that reached its zenith in his chapter on race. The recipe seemed disarmingly simple. First make your claim. When no evidence exists to support it, simply repeat the claim four times and hope the reader doesn't notice what is missing. For someone apparently so devoted to the discovery of natural truths, Galton was surprisingly quick to abandon his scientific methods when it came to the issue of race.

If the African Negro came off badly, spare a thought for the Australian Aborigine. Galton seemed reluctant to offer any judgement: 'I possess a few serviceable data about the natural capacity of the Australian, but not sufficient to induce me to invite the reader to consider them.' But despite his lack of information, he confidently asserted that: 'The Australian type is at least one grade below the African negro.'

These kinds of racial assessments were fairly typical of the time. In the eighteenth and early nineteenth centuries, Europeans had simply asserted that African blacks were inferior to whites, perhaps as a way of distancing themselves from the horrific realities of slavery. By the 1870s slavery was long gone from Britain, but the racism remained. Now the argument took on an evolutionary flavour. In an age of industry and endeavour, social progress was seen as synonymous with biological progress. The more primitive technology of blacks was taken as evidence of an underlying biological inferiority, and this kind of logic was used to rank all races on a hierarchical scale of human worth. Of course the idea was created and promoted by white Europeans, who considered themselves the supreme expression of cultural, moral, and biological achievement.

With *Hereditary Genius*, however, Galton introduced a new twist to the argument. Challenging the Victorian Briton's sense of superiority, he came up with the sensational idea that there had once been a race still higher up the evolutionary scale than upper-class Victorians. This group of people could do no wrong. They were masters of suave, sophisticated, intelligent discourse. And they built great cities, philosophies and civilisations. He was talking, of course, about the Ancient Greeks.

For some reason, Galton's critical faculties went all peculiar on the subject of the Athenians, *circa* 500 BC. Notwithstanding the great achievements made by the Greeks during this period, Galton rather naively exaggerated their merits, portraying them as a pure nation of sandal-wearing eggheads, dedicated solely to the pursuit of higher knowledge. In his own fantasy, Socrates and Phidias were the ultimate superheroes in an era of supermen: 'The millions of all Europe, breeding as they have done for the subsequent 2000 years, have never produced their equals.' Using the kind of extravagant numerical extrapolations that would make a bad economist blush, he somehow came up with the conclusion that Ancient Greeks were, on average, two grades of intelligence above Anglo-Saxons.

> This estimate, which may seem prodigious to some, is confirmed by the quick intelligence and high culture of the Athenian commonalty, before whom literary works were recited and works of art exhibited, of a far more severe character than could possibly be appreciated by the average of our race, the calibre of whose intellect is easily gauged by a glance at the contents of a railway book-stall.

Galton wasted considerable amounts of time and trees in his pursuit of the unprovable. That he persevered in such a dedicated fashion said as much about his adherence to eugenics as to scientific truth. It mattered to Galton that races differed in their innate abilities because it provided a quasi-justification for eugenic progress. The 'fact' that Anglo-Saxons were, on average, two grades above blacks showed that Anglo-Saxons had evolved to a higher level. But if Anglo-Saxons were two grades below the Ancient Greeks, then there was obviously still plenty of room for improvement.

Galton had made a convincing case that eminence, by his own definition, had a tendency to run in families. But with the results as they stood it was still impossible to distinguish between the effects of environmental and hereditary influences. For Galton, environmental explanations were clearly anathema. Human beings were not blank slates; they were already preformed and pre-destined when they came out of the womb.

I have no patience with the hypothesis occasionally expressed, and often implied, especially in tales written to teach children to be good, that babies are born pretty much alike, and that the sole agencies in creating differences between boy and boy, and man and man, are steady application and moral effort. It is in the most unqualified manner that I object to pretensions of natural equality. The experiences of the nursery, the school, the University, and of professional careers, are a chain of proofs to the contrary.

Galton did not believe that education, social class, and family influence had no impact on an individual's development, but he preferred to see these factors as an enabling, facultative force rather than a creative one. The qualities themselves, he believed, were innate, but these qualities needed nurturing if they were to find their fullness of expression. Eminence, he continually reasserted, was not something that could be acquired. It was something you were born with.

With a few notable exceptions, most of Galton's eminent men

came from the middle and upper classes. But he refused to accept the importance of a privileged upbringing in the acquisition of eminence. He did acknowledge that social factors could hinder the progress of an able man born into the lower classes. But true eminence, he asserted, would always rise to the surface, irrespective of background: 'If a man is gifted with vast intellectual ability, eagerness to work, and power of working, I cannot comprehend how such a man should be repressed.'

To reinforce his argument, Galton pointed to the United States and its more egalitarian social structure. If the English class system really did hinder the potential achievements of those from lower social ranks, then you would expect, he argued, to see a higher proportion of eminent men in America, where social hierarchies are less pronounced. To Galton, this was evidently not the case. 'America most certainly does not beat us in first-class works of literature, philosophy, or art,' he wrote.

> The higher kind of books, even of the most modern date, read in America, are principally the work of Englishmen. The Americans have an immense amount of the newspaper-article-writer, or of the member-of-congress stamp of ability; but the number of their really eminent authors is more limited even than with us.

Galton's assessment was obviously convenient to his own hereditarian argument. But when he elaborated his point, he revealed a wider, political vision: 'I argue that, if the hindrances to the rise of genius, were removed from English society as completely as they have been removed from that of America, we should not become materially richer in highly eminent men.' Galton, it seems, was more than happy with the status quo.

While he insisted that the structure of English society could not repress an intellectual giant, he also asserted that social advantage, by itself, was no passport to eminence. Having wealth, a good education, and all the right connections were no substitute for natural ability. In this sense, Galton departed somewhat from the more complacent views of the typical Victorian conservative. It was a widely held belief among the upper classes that social hierarchies reflected an underlying natural order. In other words, people belonged to the upper classes because they were naturally superior to those below them. While Galton accepted that the upper class was a repository for eminence, he did not think that the two were necessarily the same thing. Because the structure of society had some fluidity, he thought that eminent men, whatever their origins, would inevitably rise through the social ranks. It was this ongoing process of recruitment, he believed, that had made the upper classes so rich in eminent families in the first place. On the other hand, he was sometimes dismissive of those with lesser abilities, who found themselves with wealth and status purely through accident of birth. In this regard he accepted that many so-called eminent statesmen 'would have been nobodies had they been born in a lower rank of life'. Whatever prejudices he possessed in favour of the social elite, Galton's eugenics seemed to depend on the formation of a natural aristocracy rather than one built on class and status alone.

Unlike its slimline progenitor, *Hereditary Genius* succeeded in creating a stir. For a six-month period around the Christmas of 1869 the newspapers were full of reviews and comment. The *Spectator*'s review set the tone: 'His book is a very clever one, though it belongs somehow, with its shrewdness and crotchettiness [*sic*] and acute sense and absurd nonsense, to another age rather than this.' While some of the book's critics complimented Galton on his endeavour and originality, most agreed that he had overstated his hereditarian cause. *The Times* typified the mood: 'Mr Galton is a little too anxious to arrange all things in the wedding garment of his theory, and will scarcely allow them a stitch of other clothing.' Echoing these concerns, Herman Merivale, writing in the *Edinburgh Review*, added some astute observations of his own. Merivale pointed to Galton's chapter on judges. Of the 250 eminent relatives of judges, a hundred were also legal men. From this, Merivale asked rhetorically, were we to assume not only that talent was hereditary, but also the special qualities that made someone adept at law? Merivale proffered an alternative interpretation: 'The whole list has the unmistakable character of a snug little family party of jobbers, rather than that of a galaxy of genius.'

Perhaps the most perceptive comments came from the *Saturday Review*, which began with a damning, if overly harsh indictment of Galton's efforts: 'The long array of names and figures which are made to prop up the hypothesis of hereditary genius, however interesting as bits of biography, seems to us logically worth nothing.' But the reviewer followed up with an astute and very modern-sounding appraisal of the problem at hand: 'We are not to be driven, as he seems to think, to the opposite absurdity that . . . no germ of mental difference whatever accompanies the transmission of physical life. The problem for all is to distinguish this elemental germ among the mass of elements of an external kind among which it has its life and growth.' How can we tease apart and measure internal and external influences, when life requires the unified efforts of both for its proper development? This question is as pertinent today as when it was so eloquently expressed in 1869.

The scientific press greeted *Hereditary Genius* with modest praise. Writing in the newly established *Nature* magazine, Alfred Russell Wallace picked out the concluding chapters on the Ancient Greeks and eugenics as worthy of special note: '[They] stamp Mr Galton as an original thinker, as well as a forcible and eloquent writer; and his book will take rank as an important and valuable addition to the science of human nature.' But elsewhere Galton's eugenic proposals were given short shrift. The *Daily News*, for instance, pointed out some fairly basic problems that Galton had apparently overlooked: 'Unfortunately, young men will fall in love, and girls will marry them without considering the effect of the union upon the race.' The critic in *The Times*, meanwhile, suggested that Galton was missing the point entirely: 'The universal knowledge of reading, writing, and ciphering and the absence of pauperism would raise the national grade of ability far quicker and higher than any system of selected marriage.'

For all its failings, *Hereditary Genius* was an important book. The popular consensus at the time held that human heredity was a sphere of knowledge that lay beyond the reach of scientific inquiry. But Galton had shown how questions of human inheritance might be addressed. His use of family trees and statistics indicated that inroads could be made, even when the mechanism of heredity itself remained a mystery. With analysis of pedigree, Galton pioneered a technique that remains one of the most important research tools for human geneticists. Family trees are the charts of our hereditary history, the ancestral route maps that help guide the search for genes that cause human disease.

Yet for all the practical innovation contained within *Hereditary Genius*, the book's critics were spot-on in their assessments. Galton's central thesis, implicit in the book's title, was deeply flawed. With the tools at his disposal, Galton could never prove that genius was hereditary. His reluctance to acknowledge the limitations of his research suggested that, consciously or otherwise, he was working to some kind of alternative agenda. Time and again he presented his argument as straightforward biological reasoning, whereas the only thing straightforward about it was its lack of objectivity. Ultimately, he was successful in bringing the issue of hereditary determinism out into the open. But in taking such an extreme position he gave birth to an argument that has been tainted with bias and scandal ever since.

10

Rabbit Stew

Insincerity in the search after truth is one of the most degrading of sins.

Francis Galton

Galton was evidently aggrieved by the lukewarm reception granted to *Hereditary Genius*. In his autobiography, written forty years later, there are lingering signs of bitterness. '*Hereditary Genius* made its mark at the time, though subjected to much criticism, no small part of which was captious or shallow, and therefore unimportant.' There was, however, at least one opinion that he truly prized: 'The verdict which I most eagerly waited for was that of Charles Darwin, whom I ranked far above all other authorities on such a matter.'

Darwin wrote to Galton in December 1869. The letter, when it arrived, did not disappoint:

I have only read about 50 pages of your book (to Judges), but I must exhale myself, else something will go wrong in my inside. I do not think I ever in all my life read anything more interesting and original – and how well and clearly you put every point! George [his son], who has finished the book, and who expressed himself in just the same terms, tells me that the earlier chapters are nothing in interest to the later ones! It will take me some time to get to these later chapters, as it is read aloud to me by my wife, who is also much interested. You have made a convert of an opponent in one sense, for I have always maintained that, excepting fools, men did not differ much in intellect, only in zeal and hard work; and I still think this is an *eminently* important difference. I congratulate you on producing what I am convinced will prove a memorable work. I look forward with intense interest to each reading, but it sets me thinking so much that I find it very hard work; but that is wholly the fault of my brain and not of your beautifully clear style.

Until now, communication between the two men had been irregular. Ill health tended to keep Darwin away from society affairs in London, where Galton plied his trade. And although Galton made occasional visits to Darwin's home in Kent they were never great friends. Galton had great reverence for his cousin, but their personalities were not well matched. As scientists, they couldn't have been more different. Darwin worked at a pace that was slow and methodical, like the languid flow of a mighty river, while Galton preferred the current fast and turbulent. But at the start of 1870 these divergent talents were to meet in a brief and unlikely confluence. Darwin's letter to Galton, brimming with praise, signalled the start of an intense period of correspondence.

Fittingly, it was heredity that brought these cousins closer together. Galton realised that his theory of hereditary genius would always be treated suspiciously so long as there were so many loose ends and unanswered questions. And he knew that one question, above all, loomed large: the mechanism of heredity. Darwin, too, was a man in search of explanations about heredity. His entire evolutionary theory depended on the existence of heritable differences between individuals. It was the raw material from which natural selection would choose its winners and losers. Nobody disputed the fact that individuals differed from one another. What was in doubt was the material basis of this variation, and how it was transmitted from parents to offspring. Until these questions were answered, Darwin's sceptics would never be satisfied.

Today, it is difficult to appreciate the state of ignorance that surrounded heredity in the Victorian era. It may have been the age of the machine, but biologists were being held back by the limited tools at their disposal. It was not until the late nineteenth century, with improvements in microscope design and the use of chemical stains and dyes borrowed from a burgeoning textile industry, that the 'coloured body' of the chromosome slowly came into view.

Had you wandered out into an average street in 1870, and taken a straw poll of opinion on heredity, you would have returned with a bizarre mixture of old wives' tales, superstitions and a smattering of common sense. So little was known about heredity that the topic was pretty much a free-for-all, an ideal platform from which cranks, soothsayers, and even scientists could spout off whatever idea came into their head.

A common misunderstanding was that maternal impressions could affect the characteristics of an unborn child. So women in East Anglia refrained from eating strawberries during pregnancy for fear that it might cause strawberry-coloured birthmarks on their baby's skin, while pregnant Parisian women would make frequent trips to the Louvre in the hope that some of the beauty in the paintings would rub off on their children. Joseph Merrick, otherwise known as the Elephant Man, had no doubt about the importance of maternal impressions: he put his own extreme disfigurement down to the fact that his mother had been jostled and knocked over by a circus elephant during pregnancy. Merrick attributed the rarity of his condition to the rarity of elephants in rural Leicestershire.

Although the theory of maternal impressions failed to gain scientific support there was a widespread belief that characteristics acquired during adult life could rub off on a child. The idea of the 'inheritance of acquired characteristics' dated back to the work of the eighteenth-century French evolutionist, Jean Baptiste

Lamarck. Darwin himself was committed to the inheritance of acquired characteristics, and used it to explain the enlarged udders of milking cows and the small wings of domestic ducks. But for all his enthusiasm, the evidence was riddled with holes. Why did blacksmiths, for instance, who built up large muscles through their daily routine, not produce body-building babies? Why did the foreskin of circumcised fathers keep reappearing with each new generation? And why, for that matter, were the offspring of amputees not lacking in limbs? These, and dozens of other examples, made the theory seem confused and uncertain.

Some things seemed more definite. The fact that a child's characteristics were often intermediate between those of their parents lent support to the idea of bi-parental inheritance, that is, that both parents contributed equally to their offspring, via the union of sperm and egg. It also suggested that inheritance was a matter of blending, akin to the mixing of paints. But even blending could not explain the whole story. Some parental features were inherited in an all-or-nothing fashion. And sometimes a characteristic could disappear in the child, only to reappear in a grandchild or some future generation.

For Galton, this intermittent reappearance of ancestral characteristics was a particular concern. Critics had used it as ammunition in their war on *Hereditary Genius*. If genius was hereditary, they had asked, then why did it skip generations? Why were the children of talented parents not always talented themselves? Galton may have shown that talent was disproportionately common in certain families, but he still lacked any explanation as to why there was no consistency in its transmission.

In his 1868 book, *The Variation of Animals and Plants Under Domestication*, Darwin unveiled a theory of heredity that took into account all the observable facts of inheritance as he understood them. The details of the theory were necessarily speculative. But Darwin was, at least, making some conceptual steps in the right direction. Crucially, he believed that heredity came from minute particles or units within the body. The idea was very much in keeping with the age. After all, physicists and chemists had their atoms, so why couldn't biologists have something similar?

Darwin gave his hereditary theory the rather grand title of 'pangenesis', a term which accurately reflected his belief that all parts of the body contribute to the formation of a new life. He proposed that every cell within the body produced a miniature version of itself. These minute particles, which he called gem-mules, were shed from the body's cells and bundled together in the sperm and eggs. In one sense the gemmule was a conceptual precursor of the gene, in that it was a molecule that carried hereditary information. But there the similarities ended.

Darwin held out great hopes for his new theory. 'The chapter on what I call Pangenesis will be called a mad dream,' he wrote to the American botanist, Asa Gray, 'but at the bottom of my mind, I think it contains a great truth.' When pangenesis went public the scientific establishment greeted it with a cool response, just as Darwin had predicted. He confessed that his hereditary scheme was only a provisional hypothesis, but he had, at least, come up with a theory that was amenable to experiment – a fact not lost on his attentive cousin. Galton had been following developments with interest and had spotted a way of putting pangenesis to the test.

In the published account of pangenesis Darwin speculated that the gemmules 'circulate freely throughout the system'. Galton took this to mean that the gemmules are carried to the sex organs via the bloodstream. Testing the idea, therefore, seemed relatively simple. Blood could be transfused from one animal into another. If gemmules were carried in the blood, as Darwin insisted, then gemmules from the donor animal should turn up in the gonads of the recipient and manifest themselves in any subsequent offspring. In December 1869 Galton announced his intent in a letter to Darwin. 'I wonder if you could help me. I want to make some peculiar experiments that have occurred to me in breeding animals and want to procure a few couples of rabbits of marked and assured breeds.'

Fearing that his house in Rutland Gate might be overrun, Galton took his rabbits across town to London Zoo, where space and assistance were in plentiful supply. The basic plan was simple. One breed of rabbit, a silver-grey, was nominated as the recipient. In each experiment, male and female silver-greys would receive blood from another breed with different characteristics. Galton would then breed the male and female silver-greys and look for any evidence of donor characteristics in the offspring.

By March 1870 the experiments were under way. Galton could barely contain himself as he waited for the first litters to emerge. 'As soon as I know *anything* I will write instantly and first to you,' he wrote to Darwin. 'For my part, I am quite sick with expected hope and doubt.' But the early signs were not good. When the first litters were born, none of them showed any evidence of alien blood. A letter from Darwin's wife, Emma, to her daughter Henrietta summed up the mood:

> F. Galton's experiments about rabbits (viz. injecting black rabbit's blood into grey and *vice versa*) are failing, which is a dreadful disappointment to them both. F. Galton said he was quite sick with anxiety till the rabbits' *accouchements* were over, and now one naughty creature ate up her infants and the other has perfectly commonplace ones. He wishes this [experiment] to be kept quite secret, as he means to go on, and he thinks he shall be so laughed at, so don't mention . . .

In April the rabbit experiments were temporarily suspended. Galton was in Leamington, where his mother lay desperately ill. 'I am obliged to defer all this for a week or two longer,' he explained to Darwin, 'for my mother has been lying at the verge of death for a fortnight and I am wanted by her. She is now a trifle better and her illness – the result of bronchitis – may be less acute for a while and I may be able to get back to London.'

A new month seemed to bring a shift in fortunes. Back in

London, Galton was in enthusiastic voice. 'Good rabbit news!' he wrote to Darwin.

> One of the latest litters has a white forefoot . . . Yesterday I operated on 2 who are doing well to-day, and who now have 1/3 alien blood in their veins. On Saturday I hope for still greater success, and shall go on . . . until I get at least one-half alien blood. The experiment is not fair to Pangenesis until I do.

But the excitement proved short-lived. To rabbit breeders the occasional white foot was not unknown in otherwise pure breeding silver-greys. In fact it was the sporadic appearance of just this kind of ancestral character that Darwin and Galton were so keen to understand. Whatever its explanation, the white foot certainly didn't represent proof of alien gemmules at work, and it was quietly forgotten.

Correspondence broke off during the summer, when Galton and Louisa took their annual vacation, this time travelling to Grindelwald in Switzerland. But the outbreak of war between France and Prussia forced them to make an early return. By September the rabbit communiqués with Darwin had resumed.

Throughout his experiments Galton was beset with practical problems. The rabbits themselves seemed completely unfazed by the ordeal. It was the blood that was causing all the concern. At first Galton found that the blood clotted so quickly that transfusions were often impossible. So he began de-fibrinising the blood to remove its clotting potential. But by extracting the fibrin he feared that he might be removing that component of the blood that contained the gemmules. So finally he settled on direct transfusion. After anaesthetising the rabbits, he would hook up their carotid arteries with a cannula so that the blood flowed directly from one into the other. The new cross-circulation technique had no effect on the outcome. By the end of the year, Galton had bred eighty-eight rabbits in thirteen litters of which exactly none showed any alteration of breed.

In March 1871 Galton presented his results before a meeting of the Royal Society. For those in attendance the message was a stark one: 'The conclusion from this large series of experiments is not to be avoided, that the doctrine of Pangenesis, pure and simple, as I have interpreted it, is incorrect.' Darwin was not present at the meeting, nor had he received a preview of the manuscript. In the event, he was one of the last people to hear about Galton's very public dismissal of his pangenesis theory.

Suddenly, the amiable correspondence of the previous fifteen months went sour. Darwin was evidently livid and rushed off a letter to *Nature*, complaining that his ideas had been entirely misrepresented:

> Mr Galton . . . says that the gemmules are supposed 'to swarm in the blood' . . . and remarks, 'Under Mr Darwin's theory, the gemmules in each individual must, therefore, be looked upon as entozoa of his blood,' etc. Now, in the chapter on Pangenesis in my *Variation of Animals and Plants under Domestication*, I have not said one word about the blood, or about any fluid proper to any circulating system. It is, indeed, obvious that the presence of gemmules in the blood can form no necessary part of my hypothesis; for I refer in illustration of it to the lowest animals, such as the Protozoa, which do not possess blood or any vessels; and I refer to plants in which the fluid, when present in the vessels, cannot be considered as true blood.

Darwin had never speculated on the exact location of the gemmules, so it had been naive of Galton, he argued, to reject pangenesis on the basis that he couldn't find gemmules in the blood. But if Darwin really believed this then why had he taken such an interest in Galton's transfusion experiments? The two men had been in regular correspondence throughout 1870, and

Darwin knew exactly where Galton was heading. Galton may have been guilty of dismissing pangenesis prematurely. But Darwin's reaction was disingenuous and seems little more than a cover to hide his own personal disappointment. He must have realised that Galton's results sounded the death knell for pangenesis. But publicly he remained in defiant mood:

> When, therefore, Mr Galton concludes from the fact that rabbits of one variety, with a large proportion of the blood of another variety in their veins, do not produce mongrelised offspring, that the hypothesis of Pangenesis is false, it seems to me that his conclusion is a little hasty.

The following week, *Nature* offered Galton the chance to air his side of the story. Darwin's letter can't have made comfortable reading for Galton. Nobody enjoys a public dressing-down, especially from an all-time hero. To his credit, Galton remained calm, and showed that he could be diplomatic if the stakes were high enough. His reply was fair and well reasoned and he argued his case with all the panache of a great litigator.

He began his letter with a rhetorical question. If he had been mistaken to assume that gemmules circulated in the blood, then where did this mistake originate? He referred to Darwin's book, *Variation of Animals and Plants under Domestication,* and the one and only passage where Darwin had discussed the possible location of gemmules. Darwin had stated that the gemmules 'circulate freely throughout the system'. Beyond that, he had said little else. So Galton proceeded to dissect Darwin's phrase, going into great detail on the precise meaning of the word 'circulate'. 'The proper meaning of circulation is evident enough – it is a re-entering movement,' he wrote.

> Nothing can justly be said to circulate that does not return, after a while, to a former position. In a circulating library, books return and are re-issued. Coin is said to circulate,

because it comes back into the same hands in the interchange of business. A story circulates, when a person hears it repeated over and over again in society. Blood has an undoubted claim to be called a circulating fluid, and when that phrase is used, blood is always meant.

By all logical criteria, he concluded, Darwin's phrase implied that the blood might be a good place to look for gemmules. If that hadn't been Darwin's original intention, then he should have been a bit more accurate with his English. As he wrapped up his argument Galton couldn't resist making a few other suggestions where he thought the language could be improved.

With the main business taken care of Galton used the rest of his letter to rebuild battered bridges. In one of the wittiest passages he ever wrote, he proffered the hand of friendship to his revered leader. Despite their differences, Darwin was still the best. Not even a slippery argument about gemmules could alter that:

I do not much complain of having been sent on a false quest by ambiguous language, for I know how conscientious Mr Darwin is in all he writes, how difficult it is to put thoughts into accurate speech, and, again, how words have conveyed false impressions on the simplest matters from the earliest times. Nay, even in that idyllic scene which Mr Darwin has sketched of the first invention of language, awkward blunders must of necessity have often occurred. I refer to the passage in which he supposes some unusually wise ape-like animal to have first thought of imitating the growl of a beast of prey so as to indicate to his fellow-monkeys the nature of expected danger. For my part, I feel as if I had just been assisting at such a scene. As if, having heard my trusted leader utter a cry, not particularly well articulated, but to my ears more like that of a hyena than any other animal, and seeing none of my companions stir a step, I had, like a loyal member of the flock, dashed down a path of which I had happily caught sight, into the plain below, followed by the approving nods and kindly grunts of my wise and most respected chief. And I now feel, after returning from my hard expedition, full of information that the suspected danger was a mistake, for there was no sign of a hyena anywhere in the neighbourhood. I am given to understand for the first time that my leader's cry had no reference to a hyena down in the plain, but to a leopard somewhere up in the trees; his throat had been a little out of order – that was all. Well, my labour has not been in vain; it is something to have established the fact that there are no hyenas in the plain, and I think I see my way to a good position for a look out for leopards among the branches of the trees. In the meantime, Vive Pangenesis!

The charm offensive paid off. Within weeks he and Darwin had resumed their cordial correspondence. And in a perverse twist to the saga, both men then agreed to recommence the rabbit transfusion experiments, with Darwin this time taking a much more active role in the rearing of the rabbits. The experiments dragged on for another three years, although nothing ever came of them. But when a new edition of *Variation of Animals and Plants under Domestication* was published in 1875, it was clear that Darwin had taken something from the whole experience. Not only had he changed the wording of the offending passages in accordance with Galton's suggestions, but also he had added a revealing footnote. He admitted that he should have expected to find gemmules in the blood, but that it wasn't a necessary part of his pangenesis hypothesis. Galton, no doubt, was glowing.

While Galton's pursuit of the pangenesis problem seems to have been entirely honourable, he did have a lot of personal investment riding on the result. Had he found gemmules in the blood then it would have been thumbs up for his mentor. But it would have also implied a version of heredity that went completely against Galton's natural instincts.

At the centre of this hereditary conflict was the inheritance of acquired characteristics – the idea that modifications in the adult could be encoded in the germ line and passed on to the offspring. If true, it could have undermined Galton's entire hereditary programme. No longer would special talents be seen as gifts from a genetic God, but as the product of an environmentally moulded ancestry that anyone could acquire. Selective breeding might find no place in a world where everyone could grab a share of the Utopian pie. Education for all would be the rallying cry for a new wave of social idealists.

Galton's conclusion that blood was a redundant force in inheritance marked an important philosophical transition in biology. Having declared the circulatory system off limits, there was no way in which information about bodily modifications could find its way into the germ line. Expressing the seed of an idea that would later become attributed to the work of German biologist August Weismann, Galton argued, correctly, that hereditary information was passed from one generation to the next via sperm and eggs, and was more or less unaffected by changes occurring in the rest of the body. The germ line was a closed shop, and heredity, it seemed, very much an exclusive, in-house operation.

Galton further distanced himself from pangenesis by creating an entirely new hereditary jargon. Gemmules were out. Now it was all 'elements', 'germs', 'residues', and 'stirps'. This attempt to dissociate himself from Darwin worked a treat. When Galton submitted his revised hereditary theory to his cousin for comments, Darwin was baffled. 'I have read your essay with much curiosity and interest, but you probably have no idea how excessively difficult it is to understand', he wrote. 'I cannot fully grasp, only here and there conjecture, what are the points on which we differ . . . I believe (though I hope I am altogether wrong) that very few will endeavour or succeed in fathoming your meaning.'

But one person who did manage to decipher the code was

Darwin's old evolutionary ally, Alfred Russell Wallace. 'Your "Theory of Heredity"', he wrote, 'seems to me most ingenious and a decided improvement on Darwin's, as it gets over some of the great difficulties of the cumbrousness of his Pangenesis.' It was difficult to imagine a more unambiguous endorsement.

11

Question Time

The general standard of thought and morals in a mob of mediocrities
must necessarily be mediocre, and what is worse, contentedly so.
Francis Galton

In 1872 controversy seemed to shadow Galton's every move. He had shown in *Hereditary Genius* that he wasn't afraid of being outspoken. But now he seemed to be going out of his way to be awkward and confrontational. It was all rather strange behaviour from the self-confessed shy man of science. But if he was looking for an argument then that's exactly what he got. Throughout 1872 the public, press and scientific community all had a go at Galton.

The year started badly when Louisa's mother died in February, two days after Galton's fiftieth birthday. Louisa herself was not in the best of health and, judging by her 'Annual Record', she never had been. Rarely did a year go by without some gloomy reference to one kind of illness or another. The sickly theme began early in her life. Louisa was baptised at home because she was too ill to make it to the church. In 1856, three years into her marriage, she caught a fever which led to the removal of all her teeth. In 1869 digestive problems had eaten into her routine. In 1872 cholera was the ailment, Louisa remarking that she was 'very ill'.

Perhaps Galton had Louisa in mind when he wrote in *Hereditary Genius*, 'There is a frequent correlation between an unusually devout disposition and a weak constitution.' That comment, taken from his chapter on eminent divines, had gone down like a lead balloon in the religious press. But Galton was unrepentant, insisting that 'A gently complaining, and fatigued spirit, is that in which Evangelical Divines are very apt to pass their days'. Convinced of the connection, Galton illustrated his argument with pages of examples. Philip Henry, for instance, also known as 'heavenly Henry', was a vicar who sweated profusely in the pulpit, and whose devout and sedentary life was rewarded with a brain haemorrhage at the age of sixty-five. The seventeenth-century poet and rector George Herbert was a man afflicted by consumption and recurrent fever who grew more ill the more pious he became. And there was a minister called Harvey whose father tried to dissuade him from the Church for fear that his weak and puny stature would offend the congregation. He died shortly after his thirtieth birthday. These, and dozens of other tragic lives convinced Galton that religious people were, by and large, a sickly bunch with a shorter-than-average life expectancy.

With his data on divines, Galton believed that he'd stumbled across a paradox. Godliness, after all, was supposed to be good for you, not to send you to an early grave. He realised that the evidence was only circumstantial. If he wanted more conclusive proof, he would have to articulate the idea experimentally. But how do you go about putting God to the test? How do you bring the Almighty to account? He settled on a simple solution. He was going to use statistics to see if prayer was a pointless exercise.

The results never found their way into *Hereditary Genius*. Galton's original intention was to publish them as part of a separate article for the *Fortnightly Review*. But the magazine's editor rejected the paper on the grounds that it was 'too terribly conclusive and offensive not to raise a hornet's nest'. Galton had to wait three years – and a change of editor – before his article eventually made it into print.

The statistics themselves were nothing if not amusing. Galton had simply collated information on the average life expectancies for a range of different professions. Doctors, he found, could expect to live for an average of 67.07 years; lawyers fared a little worse at 66.51 years, while members of the clergy lasted a miserable 66.42 years. 'Hence the prayers of the clergy,' Galton concluded, 'for protection against the perils and dangers of the night, for protection during the day, and for recovery from sickness, appear to be futile in result.'

Having qualified his assertion with some hard figures, Galton followed up with some hard facts. If prayers really did serve any practical purpose then why did insurance companies, the bloodhounds of risk, not distinguish between the pious and the profane on their application forms? Why did doctors and nurses not advise prayer for patients? And why, for that matter, did churches have lightning conductors on top of their steeples?

While Galton denied prayer any objective purpose, he did accept that it could provide personal and psychological comfort. Indeed, he concluded his article with a touching sermon that formed a revealing insight into his own spiritual beliefs. To kick away a Christian crutch was not to be left without spiritual comfort. People could still find solace in the knowledge that they, and everything that surrounded them, were gifts of the beautiful laws of nature. They could find communion not with God, but with a transcendent universe, in all its materialistic glory. 'They know that they are descended from an endless past, that they have a brotherhood with all that is, and have each his own share of responsibility in the parentage of an endless future.' Galton was writing in the third person, but there was little doubt that he was making a personal confession. God may have gone for good, but the spirit lived on in scientific reason and his new eugenic faith.

Predictably, 'Statistical Inquiries into the Efficacy of Prayer' generated a heated correspondence when it appeared in the summer of 1872. While it was never Galton's intention to rubbish religion outright, his arrogant tone was clearly too much for some. A Christian response from an anonymous reviewer in the *Spectator* merely served to stir up the row and, within days, the *Spectator*'s

London office was overrun with letters from a nation of dis-
gruntled readers.

Charles Darwin wrote to Galton, congratulating him on the
'tremendous stir-up your excellent article on 'Prayer' has made
in England and America'. But it is unlikely that Louisa saw the
funny side. Her views on the issue are conspicuously absent from
her 'Annual Record', so one can only guess what this devoutly
religious woman was thinking. When the article was reissued as part
of a compendium of Galton's work more than a decade later it was
clear that some members of the family were still unhappy. 'I cannot
help greatly deploring what you have said on Prayer,' wrote Galton's
sister Emma. 'Whatever may be your ideas, I cannot see any reason
for publishing the fact to the World. It is a grave responsibility on
your part. I do hope in some of the later editions many of your
friends will persuade you to abstract that part from your volume.'

Galton was out of town when the prayer debate first kicked off in
August 1872. But there was no escape from controversy. He and
Louisa had gone to the annual meeting of the British Association
in Brighton, where another storm was brewing. This time, Louisa's
impressions did make it into her 'Annual Record': 'Stanley made
himself most conspicuous and obnoxious' was her terse summary
of this most sordid affair.

The Stanley in question was Henry Morton Stanley, the American journalist who had shot to worldwide fame after sniffing out
David Livingstone in Africa. Livingstone's whereabouts had been a
source of huge public interest in the early 1870s. As a lifelong
explorer and missionary, he had become an important symbol of
imperial pride. When he went missing in the late 1860s a worried
nation anxiously waited for news. Initially, the Royal Geographical
Society did nothing, believing that Livingstone could fend for
himself. But to the public, the Society's inactivity smacked of
indifference. After two years, and with still no word from Livingstone, the Society finally bowed to public pressure and arranged a
search and relief expedition, with Galton acting as Secretary of the
organising committee. What the Society didn't know was that

Stanley, working for the *New York Herald*, was already in Africa on Livingstone's trail. Stanley found Livingstone safe and well at Ujiji on the shores of Lake Tanganyika, and by the time the Society's search-and-rescue team arrived on the east African coast, Stanley was on his way home.

Arriving in Britain on 1 August 1872, Stanley was hailed as a hero, and welcomed by everyone except the Royal Geographical Society. The irritation at having been usurped by a journalist was transparent when Sir Henry Rawlinson, then President of the Society, put his own, peculiar spin on events. 'A belief seems to prevail that Mr Stanley has discovered and relieved Dr Livingstone; whereas, without any disparagement to Mr Stanley's energy, activity and loyalty, if there has been any discovery and relief it is Dr Livingstone who has discovered and relieved Mr Stanley.'

The public, however, saw right through the snub, and the Society was, once again, forced to backtrack. Rawlinson sent Stanley a belated letter of congratulations, together with an invitation to address the British Association meeting in Brighton on 16 August.

British Association regulars may have been irritated by the publicity surrounding Stanley's achievements. But Stanley's African adventures had the whole nation enthralled, and they could hardly complain when the public and press descended on Brighton in their droves. Members of parliament, bishops, and even the exiled French emperor joined the 3,000-strong crowd.

The talk was scheduled to begin at 11:00 a.m., but the auditorium was already full by 10:30. It was left to Galton, as President of the Geographical Section of the British Association, to introduce the speaker. Galton didn't talk for long. He knew little about this man of the moment so there wasn't much to say. But perhaps Stanley himself, Galton ventured, would address some of the outstanding questions regarding his origins and background.

Stanley had always maintained that he was American born and bred. But rumours were circulating that he was actually the illegitimate son of a Welsh farmer from Denbigh, who had been raised in a local workhouse. The rumour turned out to be true. Stanley was christened John Rowland, and had gone to America at the age of seventeen, taking on his adopted family's name. Stanley disowned his Welsh roots, he later claimed, for the sake of his Welsh mother, who had remarried and was still living in Denbigh.

Only Galton knows why he chose to bring up this issue in front of several thousand adoring Stanley fans. Perhaps it was all part of the Royal Geographical Society's ongoing effort to humiliate Stanley and discredit his achievements. If so, there can have been few instances where Galton misread the situation so comprehensively. Most of the crowd were not there to learn about the details of Stanley's family pedigree. They wanted to hear about his African adventures, and Stanley did not disappoint. 'I consider myself in the light of a troubadour,' he began, 'to relate to you the tale of an old man who is tramping onward to discover the source of the Nile.' He spoke at length about his journey inland to Ujiji, his meeting with Livingstone, and their exploration of the northern end of Lake Tanganyika. The press and public lapped it all up, frequently interrupting Stanley's address with enthusiastic applause and cheering.

But the style of Stanley's presentation was evidently not to everyone's tastes. It was all very well to tell a great adventure story, but for the geographic purists among the audience, the account fell short of expectations. Where were all the geographical details, the facts and figures that spelt out the course of his journey? In his own opinion, Galton told the audience, Stanley's account was nothing more than 'sensational geography'.

The British Association meeting marked the start of an ugly row that carried on for months. A week after the event, Stanley wrote to the *Daily Telegraph* expressing resentment at 'all statements that I am not what I claim to be – an American; [and] all gratuitous remarks, such as "sensationalism", as directed at me by that suave gentlemen, Mr Francis Galton'. Meanwhile, Galton and the Royal Geographical Society continued in their attempts to discredit him, insisting that both the Queen and the world at large should be properly informed about the circumstances of his birth. But the Queen showed no interest in the controversy and, for a third time, the Society had to back down. The Society eventually bestowed upon Stanley the Victoria Medal, their highest honour. Galton, however, remained impenitent. Even in his autobiography he couldn't resist one last dig at his old foe. 'Mr Stanley had other interests than geography. He was essentially a journalist aiming at producing sensational articles.'

The Stanley affair dragged on into September, but by October the storm seemed to be moving on. Galton and Louisa were in Leamington, nursing Emma back to health after an operation to remove an ovarian tumour. Calm returned to the Galton household and for a while it seemed as though things were back to normal. But the year's hostilities were not over yet. December saw the start of another big row, and this time it was Galton who was under attack.

At the centre of this latest dispute was a new book by the eminent Swiss botanist, Alphonse de Candolle. The book in question, *Histoire des Sciences et des Savants depuis Deux Siècles*, was written as a direct retort to Galton's *Hereditary Genius*. De Candolle took Galton to task for overplaying his hereditary hand, and presented his own, contrary viewpoint, which emphasised a much stronger role for environmental influences on the formation of eminent men.

De Candolle used foreign members of European scientific societies to illustrate his point, the rationale being that foreigners had to be especially distinguished to gain membership to a country's top societies. Having selected his eminent men, de Candolle then examined their historical and geographical distributions, uncovering statistics that begged serious questions of Galton's hereditary programme. The number of eminent scientists that a country produced, for instance, had no relation to the size of its population. Switzerland, for example, had 10 per cent of all the eminent scientists, but only 1 per cent of the European population. Britain and France had twice as many scientists as

you would expect on the basis of population size alone, whereas
Spain and Portugal only had about half the expected number of
eminent men. Did these figures imply that European countries
varied in their innate abilities? Perhaps they did. But if so, then
why did the percentage of eminent men in any one country
fluctuate so wildly from one century to the next. Why did Holland, for instance, produce eminent scientists at a high rate in the
eighteenth century but a low rate in the nineteenth century? To
de Candolle, the inference was obvious. The environment must
be a crucial force in the expression of human ability. Countries
differed markedly in their economy, religion, standard of living,
educational infrastructure, language, climate, and a thousand
other ways that tended to steer people towards or away from a
career in science. De Candolle was not ruling out the possibility
of hereditary variation in ability. He was simply pointing out that
the roots of human behaviour were a little more complex than
Galton wanted the world to believe.

De Candolle's book generated a heated, if good-natured, exchange of views between the two protagonists. Gloatingly, Galton
wrote to de Candolle in his fluent French informing him that
Charles Darwin's opinion 'confirms mine in every particular'. But
Galton was unaware that Darwin, caught in the middle of a
schoolboy rivalry, had already written to de Candolle, praising
the Frenchman's book with some familiar and well-chosen words:
'I have hardly read anything more original and interesting than
your treatment of the causes which favour the development of
scientific men.'

Whatever the opinion of cousin Charles, Galton realised that he
needed to make a more rounded appraisal of eminent men if he
was to circumnavigate the criticisms of de Candolle and his fellow
detractors. A more balanced perspective was needed, one that gave
equal scope to environmental and hereditary concerns. Having
already asserted that heredity was the overriding influence in the
transmission of human ability, he went back to the drawing board
to try and uncover new ways to prove his point.

The solution he chose to adopt seemed so obvious it is surprising that no one had thought of it before. But they hadn't, and so Galton found himself, once again, in the role of pioneer. To give the issue of hereditary talent a fair trial he was going to ask people awkward questions about their lives and upbringing. In short, he was going to design and distribute a psychological questionnaire. What was an extremely novel idea for the time is now a routine research tool in psychology.

Although it seemed simple enough, this was a dangerous undertaking for a diffident man like Galton. He was about to ask Charles Darwin, Herbert Spencer, Thomas Henry Huxley, James Clerk Maxwell, Joseph Hooker, and a host of other big names to confront their inner selves. With no precedent within Victorian protocol, their response was difficult to predict, and Galton was extremely apprehensive as he prepared to announce his plan:

> Though naturally very shy, I do occasional acts, like other shy persons, of an unusually bold description, and this was one. After an uneasy night, I prepared myself on the following afternoon, and not for the first time before interviews that were likely to be unpleasant . . . by dressing myself in my best clothes.
>
> I can confidently recommend this plan to shy men as giving a sensible addition to their own self-respect, and as somewhat increasing the respect of others. In this attire I went to a meeting of the Royal Society, prepared to be howled at; but no! my victims, taken as a whole, tolerated the action, and some even approved of it.

The Galton questionnaire was a massive, cumbersome beast, stretching over seven quarto pages. One hundred and ninety-two distinguished members of the Royal Society no doubt experienced a collective cringe when this weighty package landed on their doorsteps. Some of the questions were simple and required correspondingly simple answers: place and date of birth, height,

religion, and so on. But as the inquiry progressed Galton demanded more detailed answers to increasingly suggestive questions. Here, correspondents could not escape easily with yes or no
answers. This questionnaire was a gruelling, demanding journey
into the inner self. For example, it asked 'How far do your
scientific tastes appear to have been innate? Has the religion
taught in your youth had any deterrent effect on the freedom of
your researches? Independence of judgement in social, political
or religious matters?'

To Galton's delight very few of the recipients were affronted by
the exercise and over half gave full responses. The results appeared
in *English Men of Science: Their Nature and Nurture*, published in
1874. This was, in many ways, the dullest book that Galton ever
produced. While the information itself held great interest, Galton
had made little effort to condense the raw questionnaire replies
into a more readable form, so that the book amounted to little
more than a catalogue of responses. He may have been a pioneer in
the use of the psychological questionnaire, but he still had a lot to
learn about digesting and compiling the answers.

Despite its tedious and clumsy style the book did mark the
introduction of a new phrase into the biological vernacular. He had
been hunting around for a suitable expression to describe the
internal and external influences that affect human development,
before he finally settled on the now familiar formula of nature and
nurture. '[It] is a convenient jingle of words,' he explained, 'for it
separates under two distinct heads the innumerable elements of
which personality is composed. Nature is all that a man brings with
himself into the world; nurture is every influence from without that
affects him after his birth.'

Galton described *English Men of Science* as a natural history of
English scientists, a suitably broad description that allowed
him to gather together an undisciplined mix of fascinating facts
and irrelevance. As well as the now obligatory family pedigrees,
there were also all kinds of statistical titbits on the customs,
habits and origins of his scientific men. He produced a map of

Britain showing that eminent men of science rarely came from coastal towns. He found that it was unusual for scientists, unlike classical scholars, to have clergymen as fathers; that they tended to marry women of the same temperament and height; that they came from large families; that they had large heads, and that they possessed independent minds and great reserves of energy. He also dwelt on the 'fact' that scientific men tended to have lower fertility than their parents. Among his married correspondents he could find comfort, perhaps, in the statistic that one in three were childless, although the result seemed to contradict his statement ten years earlier on the sexual capabilities of eminent men.

Most of the published responses in *English Men of Science* remained anonymous, but Galton kept the completed questionnaires of all his correspondents and, even today, some of them make quite interesting reading. In answer to the question of whether he had any special talents, or strongly marked mental peculiarities, Charles Darwin wrote:

Special talents, none, except for business, as evinced by keeping accounts, being regular in correspondence, and investing money very well; very methodical in my habits. Steadiness; great curiosity about facts, and their meaning; some love of the new and marvellous.

Somewhat nervous temperament, energy of body shown by much activity, and whilst I had health, power of resisting fatigue. An early riser in the morning. Energy of mind shown by vigorous and long-continued work on the same subject, as 20 years on the *Origin of Species* and 9 years on *Cirripedia*. Memory bad for dates or learning by rote; but good in retaining a general or vague recollection of many facts . . . I suppose that I have shown originality in science, as I have made discoveries with regard to common objects.

Darwin's reply is striking in its modesty. But Thomas Henry Huxley's response to the same question is in some ways more revealing:

> Strong natural talent for mechanism, music and art in general, but all wasted and uncultivated. Believe I am reckoned a good chairman of a meeting. I always find that I acquire influence, generally more than I want, in bodies of men and that administrative and other work gravitates to my hands. Impulsive and apt to rush into all sorts of undertakings without counting cost or responsibility. Love my friends and hate my enemies cordially. Entire confidence in those whom I trust at all and much indifference towards the rest of the world. A profound religious tendency capable of fanaticism, but tempered by no less profound theological scepticism. No love of the marvellous as such, intense desire to know facts; no very intense love of my pursuits at present, but very strong affection for philosophical and social problems; strong constructive imagination; small foresight; no particular public spirit; disinterestedness arising from an entire want of care for the rewards and honours most men seek, vanity too big to be satisfied by them.

Galton must have overlooked Huxley's frank and honest confession when he wrote, 'As regards the scientific men, I find, as I had expected, vanity to be at a minimum.' He acknowledged that he'd started out with certain preconceptions. 'My bias has always been in favour of men of science, believing them to be especially manly, honest, and truthful,' and the results of the inquiry, he insisted, 'confirmed that bias'.

If there was any bias it was in Galton's tendency to fit the results of the questionnaire to his own prejudices. He may have been striving for an objective survey, but leading questions like 'Independence of judgement in social, political, or religious matters?' and 'Originality or eccentricity of character?' were only ever likely to elicit subjective answers.

The real meat of Galton's inquiry concerned the origins of the taste for science. Galton had asked his correspondents to examine the roots of their scientific interest, posing the question 'How far do your scientific tastes appear to have been innate?' From the long list of answers he made his own, not always neutral, assessment of whether each person's response indicated an overriding influence of nature or nurture. The reply 'Always fond of plants', from an anonymous botanist, for instance, was more than enough evidence for Galton to interpret the interest as innate. In the end he concluded that six out of every ten men came out on the side of nature. This still left 40 per cent who felt that their interest came from elsewhere, and Galton gave fair and full accounts of instances where family, friends, and schools provided the formative influences.

Many of the correspondents referred to their parents as an important factor behind their scientific success. Galton seemed keen to highlight the fact that the influence of the father was mentioned three times as often as that of the mother, and he used the result as an excuse to cast his expert eye over the female psyche:

> The female mind has special excellencies of a high order, and the value of its influence in various ways is one that I can never consent to underrate; but that influence is towards enthusiasm and love (as distinguished from philanthropy), not towards calm judgement, nor, inclusively, towards science. In many respects the character of scientific men is strongly anti-feminine; their mind is directed to facts and abstract theories, and not to persons or human interests. The man of science is deficient in the purely emotional element, and in the desire to influence the beliefs of others.

The fact that Galton was now considering both nature and nurture suggested that his viewpoint had mellowed appreciably since the out-and-out extremism of *Hereditary Genius*. In fact *English Men of Science* was almost completely lacking in any of the gung-ho rhetoric of his previous offerings on heredity. True, he was happy to churn out the obligatory statistics on the eminent relatives of his scientific men. Scrutinising the family pedigrees, he showed, as he had shown many times before, that his subjects had many more eminent relatives than one would expect by chance alone. But this time he seemed more than willing to admit the limitations of his data.

While *English Men of Science* contained some positive points, ultimately the book was a disappointment. Scientifically, at least, Galton had made very little progress on the nature/nurture question. With the psychological questionnaire he had pioneered another novel scientific tool, but its use had once again left him well short of the answer he so desperately yearned for. He was now more diplomatic, but there was also a tone of despondence in his words: 'The effects of education and circumstances are so interwoven with those of natural character in determining a man's position among his contemporaries, that I find it impossible to treat them wholly apart.'

His despair was short-lived. In 1875, a year later, Galton made a breakthrough that offered the real possibility of progress. A short passage in *English Men of Science* had already hinted at this new direction. 'There are twins of the same sex so alike in body and mind that not even their own mothers can distinguish them . . . This close resemblance necessarily gives way under the gradually accumulated influences of difference of nurture, but it often lasts till manhood.' Galton was slowly uncovering a new and startling revelation. Twins, he realised, were living experiments into the relative effects of nature and nurture.

Imagine, for example, Frank and Fred, a pair of happy and healthy identical twins. Suppose that, by some unusual twist of fate, Frank and Fred were separated from one another at birth. Frank fell into a life of luxury on a lavish country estate in Birmingham, while Fred was raised by a working-class family in

Moss Side, Manchester. Since they are identical twins they share the same nature. All that differs between them is their environment – their nurture.

The case of Frank and Fred offers an excellent opportunity to untangle the complex web of nature and nurture. If they turned out to be remarkably similar, despite their different environments, then it would suggest a strong influence of nature. Big differences between the twins, however, would shift the focus more towards nurture. To uncover the answer we would need to track down Frank and Fred and invite them to fill in a questionnaire, perhaps something like the one below:

	Frank	Fred	Influence of nature
Height	175 cm	170 cm	****
Weight	90 kg	80 kg	***
Intelligence	High	Low	*
Disposition	Mad	Mellow	*
Fondness for women's company	Minimal	Extravagant	*
Hobby	Collecting extinct mammals (stuffed)	Collecting toenails of extinct mammals	****
Favourite holiday destination	Sea View Hotel, Blackpool. Room 12, bed nearest the window	Sea View Hotel, Blackpool. Room 12, bed nearest the window	*****
Height of socks on shins	20.45 cm	10.85 cm	**
Pet	Iguana called Duke	Earthworm called Billy-Joe	*

Key: ***** = nature rules
      **** = mostly nature
       *** = anyone's guess
        ** = mostly nurture
         * = nurture rules

From the results of the above questionnaire we might conclude that height, weight, hobbies, and favourite holiday destination are predominantly under the control of nature, or, at least, Frank and Fred's natures. To make this nature-and-nurture investigation more convincing and statistically reassuring we would have to send out the questionnaire to many sets of identical twins. In 1875 this was more or less what Galton tried to do. Detailed questionnaires were summarily dispatched to as many twins as he could find.

Unfortunately, Galton's grand plan was scuppered by his inability to find identical twins who were separated at birth or in their childhood. Most of his replies came from pairs who had remained together until early adulthood. But so long as the evidence played into the hands of nature he couldn't resist relaying the anecdotes of the remarkable similarities. He favoured tales of mistaken identity, of coincident ideas, and shared susceptibilities. There was even the bizarre and incredible story of a man who, while holidaying in Scotland, decided to buy a set of champagne glasses as a surprise present for his twin brother. At the same time the brother, then in England, came across what he thought was an ideal gift for his twin: a set of champagne glasses.

Ignorance of genetics meant that the rationale behind Galton's investigation was not quite the same as that which informs studies of twins in the twenty-first century, but he was, at least, heading in the right direction. Today psychologists, sociologists and geneticists continue to carry out twin studies in their attempts to understand why we turn out the way that we do. They have come a long way since Galton's ground-breaking attempt to get to grips with the Franks and Freds of this world, but some things have remained the same. Galton's tendency to emphasise similarities and ignore differences in the interests of his own prejudice has been faithfully passed on to future generations of researchers. Twin studies have often been beset with accusations of bias and even fraud. Despite, or perhaps because of, Galton's original insight, they remain, at best, an ambiguous experimental tool in studies of the origins and causes of human behaviour.

With his short and already dubious track record Galton was never likely to provide an entirely balanced perspective on the issue of nature and nurture. But his work on *English Men of Science* and his investigations into twins suggested that he might be prepared to reconsider his position. *Hereditary Genius* may have been his formal declaration of intent, but even in Galton's prejudiced brain, there was still room for a modicum of scientific diplomacy.

In truth there was little option for Galton. If he was to be taken seriously as a scientist then he had to respond to criticism from the likes of Alphonse de Candolle. And having adopted a more inclusive approach to his subject he did, in fact, adjust his own point of view. Close scrutiny of the statements he was making around the mid-1870s reveal a significant shift in the tone of his opinion. He was still clinging to his vision of heredity, but now there were additional clauses that served to soften the impact of his hard-line stance. A sign of how much his position had changed can be found in a carefully worded statement written in 1875, shortly after his work on twins: 'There is no escape from the conclusion that nature prevails enormously over nurture when the differences of nurture do not exceed what is commonly to be found among persons of the same rank of society and in the same country.' Despite the cunning wordplay, this was one of Galton's more balanced statements on the whole issue of nature and nurture: given a similar set of environmental circumstances, people will still vary from one another because they have different hereditary backgrounds.

Of course there were limits to how far Galton would modify his position on the nature/nurture issue. He had opened the door an inch or two to the possibility of environmental influence, but he couldn't afford to go much further. Heredity had to be his focus. It was the very foundation on which his eugenic philosophy was built.

Eugenics itself, or 'viriculture' as Galton termed it in the 1870s, was still struggling to gain acceptance. Few people seemed to share his enthusiasm for his fictionalised Utopia. Yet two articles from 1873 showed that the idea was still very much alive in his mind. In

a letter to *The Times*, entitled 'Africa for the Chinese', he let his
thoughts run away with him when he advocated an unusual field
experiment in eugenics:

> [In Africa], as elsewhere, one population continually drives
> out another. We note how Arab, Tuarick, Fellatah, Negroes
> of uncounted varieties, Caffre and Hottentot surge and reel to
> and fro in the struggle for existence. It is into this free fight
> among all present that I wish to see a new competitor
> introduced – namely the Chinaman. The gain would be
> immense to the whole civilised world if he were to outbreed
> and finally displace the negroe . . . The magnitude of the gain
> may be partly estimated by making the converse supposition
> – namely the loss that would ensue if China were somehow to
> be depopulated and restocked by negroes.

'Hereditary Improvement', published in January, showed that
Galton was continuing to develop and refine his philosophy.
The article spelt out his entire eugenic vision and read like a
step-by-step guide to human domestication. Step one was the
establishment of a network of regional records offices to take stock
of the nation's eugenic assets:

> We want an exact stock-taking of our worth as a nation, not
> roughly clubbed together, rich and poor, in one large whole,
> but judiciously sorted, by persons who have local knowledge,
> into classes whose mode of life differs. We want to know all
> about their respective health and strength and constitutional
> vigour; to learn the amount of a day's work of men in
> different occupations; their intellectual capacity, so far as it
> can be tested at schools; the dying out of certain classes of
> families, and the rise of others; sanitary questions; and many
> other allied facts, in order to give a correct idea of the present
> worth of our race, and means of comparison some years
> hence of our general progress or retrogression.

Measuring the mental and physical faculties of the nation seemed like an onerous task. But Galton believed that individual families, small communities, and schools could be encouraged to take on the job for themselves, with assistance provided by a network of regional eugenics offices. As the information came in, local registers would be drawn up of the people blessed with the best hereditary gifts. This exercise, he believed, would in itself promote affinities among those identified as genetically gifted, resulting in more intermarriages between them. '[They] would feel themselves associates of a great guild. They would be accustomed to be treated with more respect and consideration than others whose parents were originally of the same social rank.' He even anticipated a new culture of charity geared solely towards those nominated as good breeding stock.

Galton considered two ways in which a society of supermen might develop. In the first scenario registered families would voluntarily coalesce into a tightly-knit social unit separate from mainstream society. If the community fell victim to persecution, 'as so many able sects have already been, let them take ship and emigrate and become the parents of a new state, with a glorious future'.

In the alternative scenario the state would form an integral part of any eugenic administration. Registered families would receive encouragement through state-funded endowments, while the reproductive undesirables would be expected to repress their most basic instincts. Ominous consequences awaited those who failed to conform to a celibate life. Anyone found guilty of illicit procreation would be considered 'enemies to the State, and to have forfeited all claims to kindness'. Galton used his Victorian talent for understatement but his words were chilling none the less. Was he hinting at penury, sterilisation, or even death?

He was far less ambiguous on the setting of any future society. The influence of a rural childhood shone through his insistence that human domestication was most definitely a country pursuit. Cities had sucked people out of the countryside to feed the industrial boom. To Galton they were poisonous, decadent places that provided the perfect breeding ground for all the wrong kinds of people:

> [The] ordinary struggle for existence under the bad sanitary conditions of our towns, seems to me to spoil, and not to improve our breed. It selects those who are able to withstand zymotic diseases and impure and insufficient food, but such are not necessarily foremost in the qualities which make a nation great. On the contrary, it is the classes of a coarser organisation who seem to be, on the whole, most favoured under this principle of selection, and who survive to become the parents of the next generation. Visitors to Ireland after the potato famine generally remarked that the Irish type of face seemed to have become more prognathous, that is, more like the negro in the protrusion of the lower jaw; the interpretation of which was, that the men who survived the starvation and other deadly accidents of that horrible time, were more generally of a low and coarse organisation.

Galton was committed to the idea that a healthy life meant a country life; the romantic spirit of Arcadia seemed to infuse his entire vision. In one remarkable passage he imagined a future world in which the studs and mares of his breeding programme would be adopted by wealthy landowners, for a life of bucolic bliss down on the farm:

> Queen Elizabeth gave ready promotion to well-made men, and it is no unreasonable expectation that our future land-owners may feel great pride in being surrounded by a tenantry of magnificent specimens of manhood and womanhood, mentally and physically, and that they would compete with one another to attract and locate in their neighbourhood a population of registered families.

Throughout 'Hereditary Improvement' Galton assumed a calm and commanding tone, littering his article with bold, portentous statements. 'It is the obvious course of intelligent men – and I venture to say it should be their religious duty – to advance in the direction whither Nature is determined they shall go,' he intoned solemnly, 'that is towards the improvement of their race.' Yet his breathtaking self-assurance was matched only by his manifest misunderstanding of people and their fundamental drives. Galton was advocating a society in which the dispossessed are denied even the basic right to breed. He countered the claim that his system was undemocratic by arguing that the very principles of democracy were flawed: '[Democracy's] assertion of equality is deserving of the highest admiration so far as it demands equal consideration for the feelings of all, just in the same way as their rights are equally maintained by the law. But it goes farther than this, for it asserts that men are of equal value as social units, equally capable of voting, and the rest. This feeling is undeniably wrong and cannot last.' Professing to take his lead from the natural world, Galton argued that the individual should count for nothing:

> If . . . we look around at the course of nature, one authoritative fact becomes distinctly prominent, let us make of it what we may. It is, that the life of the individual is treated as of absolutely no importance, while the race is treated as everything, Nature being wholly careless of the former except as a contributor to the maintenance and evolution of the latter.

To illustrate his idea he drew parallels with sterile worker ants, sacrificing their own procreative ambitions in the interests of the colony. But what seemed like a good analogy at the time only serves to highlight the ignorance of his reasoning. If evolution teaches us anything, it is that individuals are selfish creatures who devote their lives to getting as many of their own genes into the next generation as possible. They don't care about the greater good of the race or any other so-called lofty aims. Worker ants may *seem*

selfless but they are, in reality, only following the same evolutionary rules as everyone else. Nature, Galton always argued,
should inform the rules by which human societies are run. He
just failed to get the rules right in the first place.

The truth was that Galton's Utopia was a theoretically confused
vision of a totalitarian state. It was a stud farm for intellectuals, an
absurd world populated entirely by Newtons, Mozarts, Shake-
speares, and maybe even Galtons, an incongruous amalgam of
eggheads who were paid to have sex while women stood by on the
sidelines as little more than willing wombs. In the 1870s even the
austere Victorians found this emotionally barren vision of the
future unpalatable.

12

Vital Statistics

Until the phenomena of any branch of Knowledge have been submitted to measurement and number it cannot assume the status and dignity of a science.

Francis Galton

Louisa had been coughing up blood all night. The doctor was on his way, but it was miles to the nearest town and there was no saying when he would arrive. Sitting on the bed by her side, Galton held his wife's limp hand securely between his palms. Silence had descended on her body, and an invisible shroud seemed to glaze the surface of her porcelain skin. Not even the amber glow from a flickering candle could bring colour to her face. Galton knew she was close to death.

The air in the room felt stale and muggy. He moved away from the bed to open a window. It was still dark outside, but he could make out the rabbits in the adjoining field, already scouring the ground for breakfast. He leaned out and took a deep breath of Dorset air, picking out the country scents that filled his nostrils. When he turned back to look at Louisa, sputum was streaming over her lips.

It should have been a happy holiday with friends. In early September 1874, the Galtons had joined Sir William Grove and his family at a rented house near Blandford in Dorset. But on the fourteenth Louisa suddenly collapsed and started vomiting blood. Sir William's son took his horse and galloped several miles to summon a doctor. When he returned, with a large bag of ice, Galton and the Groves were gathered by Louisa's bedside, praying that she would make it through the night.

By morning the retching had subsided. Confined to her bed for weeks she slowly started to recover her strength. 'Thro' God's mercy,' she recalled in her 'Annual Record', 'I came back to life and felt so peaceful and happy in my quiet sick room, that it was not a time of misery. And all were so kind and good to me, and Frank especially, that I felt sustained by love.' But Louisa remained weak and frail. Whatever the cause of the illness – she put it down to a burst blood vessel – she was never the same again. A glance at her 'Annual Record' for 1874 shows a noticeable deterioration in her handwriting, a symptom of a more general physical decline that would continue for the rest of her life.

Louisa summed up 1874 as 'a year of sad memories'. It was an apt description for a period that saw not only her own serious illness in September, but the death, in February, of Galton's mother, Violetta. Galton and his mother never shared the same kind of intimacy that he established with his father. His mother always showed loving concern for her youngest son, never more so than during his second nervous breakdown in the 1860s. Any reticence in their relationship probably had its origins with Galton and his aloof attitude towards women. In his autobiography, Galton gave very little attention to Violetta. 'My mother always showed the greatest affection to me throughout her long life, which closed in 1874' was his curt summation of her ninety years on earth.

Louisa's recuperation in the autumn coincided with a decline in her husband's health. Galton began complaining of chest pains, and he was advised by his doctor to avoid eating rich and spicy foods, and to restrict his drinking to a pint of Bordeaux a day. The illness, whatever it was, proved minor. With a change of diet and a little more exercise, he was soon back to full strength.

As Galton grew older he showed no sign of slowing down. On the contrary, age seemed to accelerate his pace of work. His tally of publications for the 1860s – including books, original articles, reviews, and letters – numbered about forty. In the following decade this rose to around sixty. Now well into his fifties, he was hitting the most productive period of his career.

While heredity-related articles were starting to dominate his output, his list of publications always contained a healthy mix of the profound and the prosaic. For every 'Hereditary Improvement', for instance, there was an article like 'Bicycle Speedometer', published in an 1877 issue of *Field* magazine. The speedometer in question was one of Galton's less celebrated inventions, and consisted of nothing more than an egg-timer, which the cyclist was supposed to hold while counting the revolutions of the pedals. The number of turns in the allotted time gave the speed in miles per hour. The size of the sand glass had to be calibrated to the diameter of the bicycle wheel for the system to work properly. Perhaps it wasn't much of a surprise that it never caught on.

Galton was also a frequent contributor to the letters pages of *Nature*. He never had any shortage of things to report. A trip to the Epsom Derby in 1879, for instance, prompted a short communication entitled 'The Average Flush of Excitement'. The day was a memorable one, but not because of the horse racing. Galton's attention was evidently on other matters:

> I had taken my position not far from the starting-point, on the further side of the course, and facing the stands, which were about half a mile off, and showed a broad area of white faces. In the idle moments preceding the start I happened to scrutinise the general effect of this sheet of faces, both with the naked eye and through the opera-glass, thinking what a capital idea it afforded of the average tint of the complexion of the British upper classes. Then the start took place; the

magnificent group of horses thundered past in their fresh vigour and were soon out of sight, and there was nothing particular for me to see or do until they reappeared in the distance in front of the stands. So I again looked at the distant sheet of faces, and to my surprise found it was changed in appearance, being uniformly suffused with a strong pink tint, just as though a sun-set glow had fallen upon it. The faces being closely packed together and distant, each of them formed a mere point in the general effect. Consequently that effect was an averaged one, and owing to the consistency of all average results, it was distributed with remarkable uniformity. It faded away steadily but slowly after the race was finished.

Throughout the 1870s eugenics was still Galton's primary focus. But there was now a marked shift in the emphasis of his research. He took a holiday from the proselytising and turned instead to the empirical foundation of his plan. Before any kind of eugenic strategy could be put into practice, he needed to take stock of the nation's eugenic assets, to catalogue its range of mental and physical features. There was, Galton insisted a 'pressing necessity of obtaining a multitude of exact measurements relating to every measurable faculty of body or mind, for two generations at least, on which to theorise'. This huge collection of human data, he anticipated, would form a base line against which the success of a future selective-breeding campaign could be compared.

Schools, Galton believed, were ideal places to start collecting the kind of information he needed. They were like eugenic shops in which masters could keep regular updates of their stock. 'If a schoolmaster were now and then found capable and willing to codify in a scientific manner his large experiences of boys . . . as a naturalist would describe the fauna of some new land, what excellent psychological work would be accomplished?' In 1874, with the backing of the Anthropological Institute, Galton wrote to a number of public schools, asking for basic data on the age, height, and weight of boys. With the notable exception of Marl-borough College, the request was greeted with a large measure of indifference.

The patent lack of interest forced Galton into alternative lines of inquiry. He was always on the lookout for new ways of distin-guishing and grading people. Like many big-headed Victorians, Galton believed that skull size was a reliable indicator of intelli-gence. Since there was no objective way to measure intelligence, the claim was difficult to verify, but the general idea held great appeal for Galton. If physical features could be used as a guide to psychological ones, then it would seriously speed up his eugenic screening.

A link between the mental and the physical received widespread scientific support in the latter half of the nineteenth century. Few did more than the French anthropologist, Paul Broca, to popular-ise the idea that bigger brains – and hence bigger skulls – went hand in hand with greater intelligence. The idea fitted neatly into the whole Victorian scheme of racial hierarchy. The upper-class white male was seen as the most highly evolved of the human species, with the biggest skull and the biggest brain. To move down the racial scale was to enter a more primitive evolutionary world. In this regard, women, members of the working class, and blacks were all deemed small-brained inferiors.

Anthropologists spent an enormous amount of time trying to find characteristics that confirmed a closer association between apes and the 'lower' races of man. Years of psychological and physical profiling convinced the Italian physician, Caesar Lom-broso, that criminals belonged to one of these so-called sub-races of humanity. According to Lombroso, a criminal was 'an atavistic being who reproduces in his person the ferocious instincts of primitive humanity and the inferior animals'. Large jaws, high cheekbones, handle-shaped ears, idleness, and a love of orgies were just some of the many features that, he believed, characterised the average criminal.

In the late 1870s Edmund Du Cane, Surveyor General of

Prisons, approached Galton with the idea of doing a Lombroso-type study of British criminals. Du Cane wanted Galton to examine mug-shots of different kinds of criminals and see if he could identify and distinguish them on the basis of their facial features. Did a person's facial appearance, he asked, carry a signature of their criminal mentality?

In his 1877 address to the British Association, Galton outlined his views on criminal types:

> It is needless to enlarge on the obvious fact that many persons have become convicts who, if they had been afforded the average chances of doing well, would have lived up to a fair standard of virtue. Neither need I enlarge on the other equally obvious fact, that a very large number of men escape criminal punishment who in reality deserve it quite as much as an average convict. Making every allowance for these two elements of uncertainty, no reasonable man can entertain a doubt that the convict class includes a large proportion of consummate scoundrels, and that we are entitled to expect to find in any large body of convicts a prevalence of the truly criminal characteristics . . .

The ideal criminal, Galton believed, had three peculiarities of character: 'his conscience is almost deficient, his instincts are vicious, and his power of self-control is very weak'. The question was, did these mental characteristics manifest themselves in the physical features of his face?

Galton began his investigation with a huge stack of photographs, which he sorted into three separate piles according to the nature of the crime. There was a pile for murderers and burglars, another for forgers and fraudsters, and a third for sex offenders. Having classified his criminals, he then scoured the photographs for features that might be specific to each type.

There were certainly no obvious differences between the faces, so Galton set about looking for more subtle identifying marks.

There are, in theory, an infinite number of measurements you can take from a face. How do you go about discovering and defining the key features, if they exist at all? Galton's ingenious solution was to make composite photographs. By exposing a series of images onto a single photographic plate he was able to produce a pictorial average of many criminal faces. The overall effect was to dissolve away individual differences and emphasise similarities. He could then compare the composites for each of the three categories of criminal and look for differences between them.

While his photographic technique worked perfectly, the composites turned out to be indistinguishable from one another. In fact the composite criminal was a picture of social respectability. If there was such a thing as the criminal mentality, it certainly didn't bear a recognisable human face.

Galton had always believed that a person's mentality left its mark on their physical features. But the results of his photographic study now seriously challenged that faith, and he was forced to rethink his position. He realised that if he wanted to learn more about mental faculties then he would have to approach the subject head-on. Galton's primary objective was to take stock of the mental differences between people. But in his effort to uncover whether the mind could submit to measurement, he slipped into the still relatively deserted field of human psychology.

The manner in which Galton confronted the study of human psychology typified his entire scientific philosophy. It is not clear whether he was aware of the work of Wilhelm Wundt and the emerging school of experimental psychology in Germany. If he was, it probably would have made little difference. With his idiosyncratic blend of naivety and arrogance, Galton simply advanced on the mind from first principles, creating his own school of experimental psychology from scratch.

Like Wundt before him, Galton believed that introspective self-analysis might provide fruitful insights into the mechanisms of the mind. Using your own mind to monitor itself is an extremely difficult technique to perfect, a bit like playing drums in one tempo while singing in another. With a bit of practice, however, Galton thought that he had it licked. 'My method consists in allowing the mind to play freely for a very brief period, until a couple or so of ideas have passed through it, and then while the traces or echoes of those ideas are still lingering in the brain, to turn the attention upon them with a sudden and complete reawakening.'

Galton test-drove his new technique with a gentle stroll down Pall Mall. Each time an object caught his attention he made a mental note of any ideas that the object provoked. Later on, when he came to scrutinise these alliances, he was astonished by the depth of the mental links between them. 'I am sure, that samples of my whole life had passed before me, that many bygone incidents, which I never suspected to form part of my stock of thoughts, had been glanced at as objects too familiar to awaken the attention.' It was a simple but eloquent demonstration of the subliminal nature of cognition. Beneath the surface of humdrum, superficial experience lay another, more curious world. And Galton showed that, with a bit of mental effort, it might be possible to go exploring.

Keen to tap into these rich mental seams, he hit upon the idea of the word-association experiment. From a dictionary he took seventy-five words beginning with the letter 'a' and wrote each one down on a separate card. This stack of cards represented his set of stimulus words. Holding a stopwatch, he would then pick a card at random from the pile, look at the stimulus word and record the time taken for two associated ideas to come into his head. If his mind was still a blank after four seconds he would move on to the next word, and continue until all had been tested.

The credibility of the experiment depended on the mind paying equal attention to each of the stimulus words. But the mind soon tires of these four-second reflections, and Galton's main problem was overcoming the profound desire to switch off. Yet he not only succeeded in wading through all the cards, he insisted on repeating the entire experiment another three times.

The four experimental trials produced a collection of 505

associations, and Galton immediately set to work on looking for different ways of classifying them. At one point he arranged his associations according to the time of his life to which he believed they were related, and found that the majority of associations came from his boyhood and early youth. The result prompted Galton, in a rare moment of weakness, to raise the national flag for nurture. Education, he asserted, was the crucial factor in fixing these early associations.

The word-association experiments offered all kinds of tantalising insights. Galton was convinced that reaction time to stimulus words, particularly the more abstract ones, was a good indicator of intellect. 'Nothing is a surer sign of high intellectual capacity than the power of quickly seizing and easily manipulating ideas of a very abstract nature.'

Galton's efforts to measure the mind were hugely prescient. Twenty years before Freud's psychoanalysis, Galton was already using word association to gain glimpses of his own subconscious. Galton, however, wasn't giving away any details of what he was finding there. '[Associations] lay bare the foundations of a man's thoughts with curious distinctness, and exhibit his mental anatomy with more vividness and truth than he would probably care to publish to the world,' he wrote.

He may have been shocked by the contents of his own mind, but it didn't deter him from further exploration. On the contrary, he wanted to raise the stakes even higher and experience for himself what it was like to be insane. It was, admittedly, a short mental leap to make, but a dangerous one none the less. What if his self-administered techniques were irreversible? Galton didn't seem to care.

'The method tried', he explained, 'was to invest everything I met, whether human, animal, or inanimate, with the imaginary attributes of a spy.'

Having arranged plans, I started on my morning's walk from Rutland Gate, and found the experiment only too successful.

By the time I had walked one and a half miles, and reached
the cab-stand in Piccadilly at the east end of the Green Park,
every horse on the stand seemed watching me, either with
pricked ears or disguising its espionage. Hours passed before
this uncanny sensation wore off, and I feel that I could only
too easily re-establish it.

Fresh from plundering one element of his psyche Galton moved
promptly on to another. 'I had visited a large collection of idols
gathered by missionaries from many lands, and wondered how
each of those absurd and ill-made monstrosities could have obtained the hold it had over the imaginations of its worshippers. I
wished, if possible, to enter into those feelings.' So Galton found
an idol of his own – a comic picture of Punch – and dedicated all
his energies to its worship.

I addressed it with much quasi-reverence as possessing a
mighty power to reward or punish the behaviour of men
towards it, and found little difficulty in ignoring the impos-
sibilities of what I professed. The experiment gradually succeeded; I began to feel and long retained for the picture a large
share of the feelings that a barbarian entertains towards his
idol, and learnt to appreciate the enormous potency they
might have over him.

Having vividly illustrated the malleability of his own mind, he
turned away from these introspective inquiries and went exploring
the minds of others. The success of his word-association experiments had prompted thoughts of memory and mental imagery in
more general terms. He considered his own abilities in this area to
be relatively poor. He had difficulty recalling visual images with any
kind of detail. But was it the same for everyone, or did people vary
in their powers of visualisation?

In November 1879 Galton came up with a questionnaire,
'Questions on the Faculty of Visualising', designed to test the

visual acuity of the mind's eye. Each correspondent was asked to picture an image in their mind – Galton suggested that morning's breakfast table – and then give information on the details of the mental image, in terms of the intensity of its illumination, the colouring, definition, and field of view. The questionnaire then broadened into more general inquiries about the correspondent's ability to visualise people, places, numbers, music, and all kinds of other objects and activities.

Galton's scientific friends were the first to fill in their questionnaires, and he was amazed by the consistency of their responses. The comments of one anonymous correspondent were typical. 'To my consciousness there is almost no association of memory with objective visual impressions. I recollect the breakfast-table, but do not see it.' With few exceptions, his men of science were almost blind to the images of the mind. For some the concept was so alien that they even doubted the sincerity of Galton's questionnaire. What were these vivid powers of visualisation to which he was alluding?

The answer came when Galton had gathered up his returns from a wider sample of society. Only then did the huge variety in the powers of visualisation become clear. The scientists represented one end of the spectrum. At the other were those who could visualise an object with such clarity that it could have been sitting right in front of their eyes. The powers of one correspondent were so strong that he felt dazzled whenever he conjured up a mental image of the sun.

The results of the survey left Galton with some explaining to do. It seemed difficult to imagine a situation where a strong visualising sense could be anything other than beneficial to intellectual thought. At the very least, engineers, artists, and architects would surely gain an advantage by thinking through problems in their mind before putting them into practice. But scientists, an integral part of Galton's eminent elite, were disproportionately poor visualisers. Did this mean, therefore, that they were all dim? Not at all, Galton insisted. Scientists and philosophers had, necessarily, to deal with abstract ideas. In these circumstances vivid mental pictures would only get in the way. 'My own conclusion is, that an over-ready perception of sharp mental pictures is antagonistic to the acquirement of habits of highly-generalised and abstract thought.' But Galton was clearly uncomfortable with his assessment because he went on to add his own speculative safety net: 'The highest minds are probably those in which it is not lost, but subordinated, and is ready for use on suitable occasions.'

While questions could be raised over the scientific rigour of Galton's questionnaire – he was, after all, dealing only in subjective impressions and interpretations – his overall conclusion seemed incontrovertible. People varied enormously in their visualising powers. This variety, Galton believed, was manifest not only within populations, but also between populations. The French, he contended, were experts in the field of visual imagery. Their genius for military strategy, their ability to organise complex ceremonial events and fêtes, even the language itself was testimony to their supreme visualising powers. The French phrase 'figurez-vous', for instance, was far more vivid and definite than the English equivalent of 'imagine'. Galton's bonhomie even extended to the 'commonly despised' but 'much underrated' Bushmen of South Africa. The remarkable attention to detail shown in the Bushmen's cave paintings was a sure sign, Galton believed, of a strong visual memory.

The breakfast-table questionnaire was a huge success for Galton. It transformed people's perceptions of the mind and became the standard format by which future psychologists would approach studies of mental imagery. It was, moreover, a reassuring study for those whose own mental experiences left them feeling isolated and alone. Hitherto private and solitary sensations were brought out of the mental closet and made public, exposing the huge variation that exists in the imaginative experience.

Visual imagery, Galton discovered, could manifest itself in a multitude of different ways. He uncovered tremendous variety, for instance, in the way in which people perceive numbers. The sound of the number six, for example, could invoke a whole range of responses depending on the individual. Some people might hear and digest the number without any recourse to visual imagery. Others might picture an image of a six in their heads. While at the extreme end of the mental spectrum were those who perceived what Galton called 'number forms'. These individuals worked with a mental canvas on which numbers occupied specific locations within a strict diagrammatic arrangement. Each person's number form had a style and complexity that was uniquely their own.

Perhaps the most bizarre examples of visual imagery came from people who reported mild forms of synaesthesia, whereby specific numbers, letters, or words prompted vivid sensations of colour. One correspondent associated a different colour with each day of the week. Wednesday, for instance, was an 'oval flash of yellow emerald green', while Friday was a 'dull yellow smudge'. Another correspondent, a Mrs H., visualised the vowels in distinctive hues. 'A' was 'pure white, and like china in texture', while 'O' was 'the colour of deep water seen through thick clear ice'. Mrs H. considered consonants to be pretty much colour-free although she did feel that there was 'some blackness about M'. What made the peculiarities of Mrs H. especially interesting was that both her daughters also felt the chromatic force of the vowels, although their colour schemes were different from those of their mother. Galton made the most of this particular ancestral association. Psychological phenomena were interesting in their own right, but a hint of heredity always added an extra twist.

If the public wanted to find out more about the colourful life of Mrs H., and others like her, then they could buy Galton's latest book, *Inquiries into Human Faculty*. First published in 1883, *Inquiries* rounded up much of the work he had done since the publication of *Hereditary Genius* in 1869. The book read as a kind of natural history of humanity, Galton style. So while the text essentially told a story of the psychological and physical features of man, it also had a peculiar tendency to wander off in unexpected directions. Galton explained, for instance, all about the different properties of raw and boiled eggs when spun on a table top. He confessed his extreme fear of snakes and then added, somewhat needlessly, that this phobia was not universal among the English. And he asserted that Bern appeared to have more large dogs lying idly about the streets than any other city in Europe.

Animals were one of the book's recurring themes. In a section on the senses Galton recounted his efforts to give hearing tests to the occupants of London Zoo. He wanted to know which of the various animals had the ability to hear high-pitched notes. To help him in his evaluation he had come up with a special apparatus, custom-made for the occasion:

> I contrived a hollow cane made like a walking stick, having a removable whistle at its lower end, with an exposed india-rubber tube under its curved handle. Whenever I squeezed the tube against the handle, air was pushed through the whistle. I tried it at nearly all the cages in the Zoological Gardens, but with little result of interest, except that it certainly annoyed some of the lions.

On page 104 of *Inquiries* Galton enlightened his readers with the news that, in his mind, even numbers were intrinsically male. To those who had followed Galton's career with any interest this wasn't actually news at all. Females were always the odd ones out in Galton's book. Discussing the relative sensitivity of the two sexes he offered a taste of his even-handedness:

> I found as a rule that men have more delicate powers of discrimination than women, and the business experience of life seems to confirm this view. The tuners of pianofortes are men, and so I understand are the tasters of tea and wine, the sorters of wool, and the like. These latter occupations are well salaried, because it is of the first moment to the merchant that he should be rightly advised on the real value of what he is about to purchase or to sell. If the sensitivity of women were

superior to that of men, the self-interest of merchants would lead to their being always employed; but as the reverse is the case, the opposite supposition is likely to be the true one.

Ladies rarely distinguish the merits of wine at the dinner-table, and though custom allows them to preside at the breakfast-table, men think them on the whole to be far from successful makers of tea and coffee.

Inquiries gave Galton the perfect opportunity to serve up his forth-right opinions on women. A section on the innate character of females exposed the bare bones of his beliefs. 'Coyness and caprice have . . . become a heritage of the sex, together with a cohort of allied weaknesses and petty deceits, that men have come to think venial and even amiable in women, but which they would not tolerate among themselves.' This was the 1880s; in literature and in real life women were slowly emerging from their male shadows. But Galton never accepted sexual equality. His response to the women's movement was to join the Anti-Suffrage Society. He acknowledged that intelligent women would be highly valued in any future eugenic community. But Galton always thought in terms of women's wombs rather than their brains.

The book's apparent mishmash of topics reflected the fact that Galton's train of thought was sometimes guided more by the love of pure measurement than the pursuit of a single scientific goal. In a section on human features, for instance, he talked about our ability to distinguish a familiar face among a crowd of strangers. Galton had recently had his portrait painted, and he explained how he had used the experience to try and gain some kind of numerical grasp on this remarkable discriminatory power. Sitting in front of the artist for a total of forty-five hours, Galton estimated that the artist used about twenty-four thousand brush strokes. It was a huge number and testimony, he argued, to human beings' acute perception of facial detail.

Inquiries contained so much food for thought that the critics were never going to go hungry. There was near-universal praise for his psychological work, for his originality and ingenuity in applying statistical methods to a subject previously considered beyond the realm of scientific inquiry. The *Scotsman*, however, seemed keen to put things into perspective: this was clearly a book to be admired rather than loved. 'Every one acquainted with Mr Galton's writings, especially his *Hereditary Genius*, will know what to expect in any book from him – laborious accumulation and sifting of facts, and acute speculation based upon them. Without being a master in science, he is one of its most useful and valuable servants, and . . . he supplies much material for larger and brilliant minds to work upon.'

To many readers, *Inquiries into Human Faculty* was a confused and loose collection of facts in search of a unifying theme. As one critic commented, 'So varied is Mr Galton's matter that the reviewer pants after him in vain.' But the author did have a rationale. To Galton, at least, there was a thread that tied together his many disparate lines of inquiry. It was, by now, an old and familiar story and it was spelt out eloquently on page one:

My general object has been to take note of the varied hereditary faculties of different men, and of the great differences in different families and races, to learn how far history may have shown the practicability of supplanting inefficient human stock by better strains, and to consider whether it might not be our duty to do so by such efforts as may be reasonable, thus exerting ourselves to further the ends of evolution more rapidly and with less distress than if events were left to their own course.

Inquiries into Human Faculty, therefore, was a companion volume to its predecessor, *Hereditary Genius*. It was an anthropological snapshot of society, a database of humanity, a reconnaissance mission for eugenic strategists.

Galton used *Inquiries* to reiterate his grand experiment in human cultivation and re-launch his social philosophy under a new banner. He had never been happy with his original term of 'viriculture', and had been hunting around for something more suitable. Now, at last, he had a name with which he was comfortable. 'Eugenics', the term he coined, came from the Greek word 'eugenes', meaning 'good in stock' or 'hereditarily endowed with noble qualities'.

But if the comments of the critics could be taken as a representative sample of public opinion, there still seemed to be little sympathy for Galton's policy of social change. Most agreed that the biggest barrier standing in the way of eugenics was human nature itself. 'The many philosophical suggestions made for improving the human race', wrote the reviewer in the *Guardian*, 'have, as even Mr Galton must admit, proved futile, because they have left out of account the most important factor of all, the influence of human will, of taste, of passion, of prejudice, of caprice.' The reviewer then added a definitive and telling remark, which must have cut Galton to the core: 'To influence the future of humanity has been the desire and the effort of the best men who have lived before and since the time of Socrates . . . Every father capable of reflection has not only hoped, but has tried to do so.'

13

The Gravity of Numbers

I know of scarcely anything so apt to impress the imagination as the wonderful form of cosmic order expressed by the 'law of error'. A savage, if he could understand it, would worship it as a God.

Francis Galton

In May 1884 Galton travelled to Cambridge University to deliver the prestigious Rede Lecture. He'd been feeling extremely anxious at the prospect of a return to his old stamping ground. Cambridge, after all, was the altar on which his intellect and health had been humbled over forty years earlier. It wouldn't be easy, stepping into the spotlight, to be displayed and scrutinised.

In the event, Galton's address turned into something of a double act. When he wasn't glancing down at his notes, he had his eye on Louisa, his companion in arms, sitting somewhere towards the back of the audience. Husband and wife had rehearsed a simple signalling system to ensure that Galton's talk ran smoothly. If the volume of his speech strayed from its ideal setting, Louisa would raise her left arm to signal more or less amplification. Should the tempo of the talk deviate from its predetermined pace, then her right arm would swing into action.

The talk, entitled 'Measurement of Character', dwelt on the possibility of reducing human personality and temperament to a measurable form:

I do not plead guilty to taking a shallow view of human nature, when I propose to apply, as it were, a footrule to its heights and depths. The powers of man are finite, and if finite they are not too large for measurement . . . Examiners are not I believe much stricken with the sense of awe and infinitude when they apply their footrules to the intellectual per-formances of the candidates they examine; neither do I see any reason why we should be awed at the thought of examin-ing our fellow creatures as best we may, in respect to other faculties than intellect. On the contrary, I think it anomalous that the art of measuring intellectual faculties should have become highly developed, while that of dealing with other qualities should have been little practised or even considered.

Nothing in human nature, he suggested, was indeterminate. Any-thing and everything could be measured. To illustrate his point he told the audience that he had a pneumo-cardiograph – a device for measuring the heart's activity – concealed beneath his jacket. The pneumo-cardiograph was not plugged in; it was for demonstration purposes only. But had there been a man crouched under the table in front of him, recording the pneumo-cardiograph's output, he could have obtained a numerical assessment of Galton's emotional state.

The pneumo-cardiograph was a good way of gauging an indi-vidual's sensitivity to a stressful situation. But there were many other aspects of human character, Galton argued, that were amenable to experiment. A simple question about the state of the weather, for instance, would quickly sort out the optimists from the pessimists.

Not all of Galton's suggestions, however, were so clear-cut. A peculiar experiment carried out at his home in Rutland Gate seemed to reveal more about his own character than those of the unsuspecting dinner guests he was struggling to measure:

The poetical metaphors of ordinary language suggest many possibilities of measurement. Thus when two persons have an 'inclination' to one another, they visibly incline or slope together when sitting side by side, as at a dinner-table, and they then throw the stress of their weights on the near legs of their chairs. It does not require much ingenuity to arrange a pressure gauge with an index and dial to indicate changes in stress, but it is difficult to devise an arrangement that shall fulfil the threefold condition of being effective, not attracting notice, and being applicable to ordinary furniture.

Difficulties with his dining-room chairs could do nothing to quell Galton's insatiable appetite for numbers. Unshrinking in his faith in eugenics, he had emerged from *Inquiries into Human Faculty* in measuring mood. There was still much groundwork to be done. Whatever the critics said, it was only a matter of time, he believed, before popular opinion woke up to the wisdom of eugenics. And he had to make sure he was ready when it did.

In 1884 he launched a new eugenic initiative. He wanted records of family histories and lots of them. He needed to know all the nitty-gritty details – not just of the individual, but of their parents, grandparents, and great-grandparents. He wanted information on height, appearance, energy, occupation, health, temperament, hobbies, longevity, and intelligence of every family member. And if there was a long history of mental or physical illness in the family, then he wanted to know all about that too.

The explanation was simple. Eugenic selection could not be based on competitive examination alone. An exam might reveal the health, character, and intellect of an individual in their youth. But hereditary characteristics that manifested themselves later in life would escape detection. Any assessment of eugenic merit, therefore, had to take account of ancestry.

Galton hoped to encourage people to take on the responsibility of keeping records for themselves. Passed on to subsequent generations these records would form a unique heirloom, an ancestral chronicle to keep alongside the family photo album. But like the school masters he had tried to coax ten years earlier, the public didn't share his enthusiasm. Even when he dangled the carrot of a cash prize to anyone brave enough to fill in one his epic questionnaires, barely one hundred people responded. Such reticence left him both angry and bemused:

> It seems to me ignoble that a man should be such a coward as to hesitate to inform himself fully of his hereditary liabilities, and unfair that a parent should deliberately refuse to register such family hereditary facts as may serve to direct the future of his children, and which they may hereafter be very desirous of knowing. Parents may refrain from doing so through kind motives; but there is no real kindness in the end.

With his progress blocked by public disinterest, Galton decided that a complete change of strategy was required. If the people wouldn't come to him then he would go to the people. In 1884 he found the ideal opportunity to bring the masses on board his anthropometric adventure.

That spring the International Health Exhibition opened in South Kensington, only a stone's throw from Galton's home. The exhibition was a celebration of everything and anything remotely associated with health at home and in the work place. There were lavish displays of fabrics and foods, costumes and cosmetics. Furniture exhibits rubbed up against demonstrations of new household appliances, construction techniques, and sanitary systems. It was a truly global spectacle, with products from all over the world. Spread over acres of galleries, the exhibition attracted thousands of visitors. The more inquisitive ones made their way to the South Gallery, where they could line themselves up to be measured in Francis Galton's anthropometric laboratory.

The laboratory itself was a narrow corridor, six feet wide by thirty-six feet long, and cordoned off from the rest of the gallery by a lattice fence, which allowed passers-by to glimpse what was going on inside. On payment of a three-pence admission fee to the doorkeeper, visitors would enter the laboratory at one end of the corridor, be led through a series of tests by a superintendent, and then exit at the other end. Galton was usually on hand to assist as measurements were made of sitting and standing height, arm-span; weight; pulling, squeezing and punching power; breathing capacity; reaction time; hearing and eyesight; colour perception, and judgement of length. At the end of the examination, each person was given a card of their results, while Galton kept a duplicate for his records.

Dealing directly with the public brought its discomforts for Galton. Too embarrassed to ask people to remove their shoes when measuring their height, he would first measure the height of their shoes' heel and then subtract it from the total. Head mea-surements were also out of the question. 'I feared it would be troublesome to perform on most women on account of their bonnets, and the bulk of their hair, and that it would lead to objections and difficulties.'

Prior to the exhibition, Galton had circulated a letter among British psychologists, appealing for information and advice on what kind of measuring apparatus he should use. His request drew a muted response, largely because the devices he needed did not yet exist. It was left to Galton to build many of them himself. He complained that it was 'by no means easy to select suitable instruments for such a purpose'.

They must be strong, easily legible, and very simple, the stupidity and wrong-headedness of many men and women being so great as to be scarcely credible. I used at first the instrument commonly employed for testing the force of a blow. It was a stout deal rod running freely in a tube, with a buffer at one end to be hit with the fist and pressing against a spring at the other. An index was pushed by the rod as far as it entered the tube in opposition to the spring. I found no difficulty whatever in testing myself with it, but before long a man had punched it so much on one side, instead of hitting straight out, that he broke the stout deal rod. It was replaced by an oaken one, but this too was broken, and some wrists were sprained.

Galton's first anthropometric laboratory, at the International
Health Exhibition in London

Despite its hazards, Galton's anthropometric laboratory was a great success. By the time the Health Exhibition closed its doors in 1885, almost 10,000 people had been measured. With over 150,000 measurements recorded it would be a large statistical meal to digest, and Galton gloated over it in eager anticipation.

By the late nineteenth century the word 'statistics' was no longer synonymous simply with numbers. A mathematical framework was emerging to help interpret and describe those numbers. Probability tests evolved to assess population parameters and enable distinctions to be drawn between them. Gradually, the collection of numbers became the study of numbers, and today statistics is a subject in its own right.

As an experimental tool statistics has assumed immense power in all areas of modern science. It has become the great arbiter of all that is scientifically true, a mathematical guide to what is likely and what is not. Publicly, its name is rarely spoken, but its presence is ubiquitous, lurking beneath the surface of everyday 'facts' of life. Consider the well-established links between smoking and cancer, classroom size and exam results; HIV and AIDS; interest rates and inflation; greenhouse gases and global warming. It is statistics that has enabled us to make these connections.

For 2,000 years science existed without the aid of statistics. Aristotle, Newton, and Copernicus all managed to get through life without it. But they, and others like them, were necessarily working within a much narrower scientific philosophy. Mathematical descriptions of the physical universe began with deterministic laws of simple cause and effect. If an apple dropped from a tree, there was never any doubt over which direction it was going to fall. Science doesn't need statistics so long as it operates in the realm of certainty. But biological phenomena are rarely so straightforward. More often than not, individual events have multiple causes. We know, for example, that smoking is an important cause of cancer, but it is not the only one. Before statistics it was impossible to separate out and quantify many disparate causes. Statistics, therefore, was a door to a whole new world of inquiry, a universal mathematical language that displaced ambiguous words like 'seldom', 'usually', and 'often' from the scientific vocabulary. It revolutionised the scope and ambition of scientific endeavour, enabling subjects like biology, sociology, and psychology to flourish.

Galton was a key player in that transformation. With his predilection for measurement, he was naturally predisposed to a science of numbers. He never consciously set out to be a statistical pioneer, but he realised that if he wanted to make progress in heredity then he would have to make progress in statistics.

Some people hate the very name of statistics, but I find them full of beauty and interest. Whenever they are not brutalised, but delicately handled by the higher methods, and are warily interpreted, their power of dealing with complicated phenomena is extraordinary. They are the only tools by which an opening can be cut through the formidable thicket of difficulties that bars the path of those who pursue the Science of man.

Ever since the 1869 publication of *Hereditary Genius* the critics' voices had been ringing in Galton's ears. If heredity was so integral to human nature, they asked, then why was it so unpredictable? Why were the children of able parents, for instance, not always of high ability themselves? The solution, Galton believed, was a statistical one. If heredity was fundamentally a matter of probability then its laws could never be deduced from individual cases. True patterns would only emerge with a significant change of scale.

Galton had already convinced himself that the bell curve offered the ideal description of the way in which many human characteristics, like height, weight, and intelligence, were distributed in a population. With the hereditary question on his mind, he now wanted to examine how the distribution of heights in one generation compared with the distribution of heights in the next.

Galton's anthropometric laboratory had furnished him with just the kind of information he needed – the heights of 205 sets of parents and their adult children. To get a visual feel for how the heights of parents and offspring compared with one another he plotted his points out on a graph. As expected, the plot showed a positive association between the two: the taller the parents, the taller the children. But the association wasn't perfect. Children tended to be less extreme than their parents. Taller than average parents tended to produce children who were shorter than them, while shorter than average parents tended to have children who were taller than them.

Keen to quantify the strength of the association, Galton took his pencil and drew the best-fitting straight line through his points. A few seconds later he measured the gradient of the line. It was two-thirds. It was all pretty simple stuff, but no one had thought of it before. Galton had just provided the first numerical evaluation of the strength of the relationship between two variables.

What did this magical number of two-thirds actually mean? In practice, it meant that for every inch increase in the average height of the parents, there was only a two-thirds of an inch increase in the average height of the children. Or, put another way, children were only two-thirds as extreme as their parents. Because of this tendency for the children's heights to revert or 'regress' back towards the population average, Galton decided to give his new statistical invention the name 'regression'.

Galton arrived at the idea of regression with his work on human heights, but there seemed no reason why his measure of associa-tion could not be applied to any pair of variables. There was one small problem, however. The regression coefficient works fine provided that the two variables are analogous. In the case of Galton's human data, he was comparing the heights of parents with the heights of their children. But with many common asso-ciations – number of cars on the road and level of air pollution; winter temperature and household fuel consumption; rainfall and crop yield – you are not comparing like with like. In these instances the regression coefficient makes no sense as a measure of association. But Galton discovered that if he transformed the measurements into standardised units he obtained a measure of association that worked in all contexts. Like the regression coefficient, the so-called correlation coefficient is a numerical guide to the strength of association between two variables.

Of course correlation is not the same thing as causation. Just because two variables are highly associated with one another doesn't mean that the one causes the other. But even if a correlation is not synonymous with a cause, it is still an important first step in any scientific investigation, a vital signpost on the road to causation.

Regression and correlation were two of Galton's greatest gifts to science. Together, they gave science a huge creative boost, opening up whole new avenues of investigation. For life scientists in particular, struggling to get to grips with the complex and multi-layered roots of biological, social, and psychological phenomena, here was a numerical tool that could guide them through hitherto unexplored and inaccessible terrain. Galton's statistical insights were like a beacon in the fog, illuminating a path to a new kind of clarity.

Paradoxically, Galton's discovery caused confusion in his own mind. With the regression coefficient of human height, he had, in effect, come up with a real insight into the relative influence of nature and nature. Unfortunately, he was too blinded by his own prejudice to see it.

Put simply, all human variation can be due to differences in heredity, environment, or a combination of the two. Think about height, for instance. Some of the variation in height is due to the different genes we inherit, and some is due to differences in childhood nutrition and other environmental factors. To work out the relative contribution of these two factors, geneticists today do exactly what Galton did; they plot the average parental values for a given characteristic against those of their offspring. If all variation is due solely to genetic factors, then you would expect to find a perfect association between parents and offspring. If, at the other extreme, all variation in a characteristic has an environmental origin, then there will be no correspondence between the parents and the offspring. So the regression coefficient measures how much of the variation is due to genes. Today, it's what geneticists call 'heritability'.

Galton didn't really appreciate any of this; he was too busy pursuing dead ends. But the idea can still be illustrated using his own data on human heights. Galton found a regression coefficient of two-thirds, or 0.66. In other words, 66 per cent of all the variation in human height was due to hereditary factors, while the rest, 34 per cent, was environmental.

At first sight it looks as though heritability could have been the answer to all of Galton's prayers, a statistical method to put his theory of hereditary genius to the test. Even today, geneticists and social scientists get quite excited about heritability estimates. High heritabilities of IQ, for instance, are often quoted in support of the view that human intelligence is primarily an innate, inborn quality. But heritabilities are not absolute answers to the relative influence of nature and nurture. They only make sense within the context of a specific population in a specific environment. Imagine, for instance, that one hundred children are offered an identical education. With a uniform environment, the children might show a high heritability of IQ simply because genes are the only thing causing variation in intelligence. But if each of these children had been educated individually and independently, more of the variation in intelligence might be attributable to environmental causes. Even though these are the same children with the same genes, the *fraction* of the total variation in intelligence due to genes, and hence the heritability, has diminished.

Despite their drawbacks, heritabilities do have their uses. For animal and plant breeders, in particular, they are vital pieces of information. The success of any selective breeding depends on the amount of genetic variation in a population. Estimates of heritability, therefore, enable breeders to predict how a population will respond to a programme of artificial selection.

Heritability would have been a handy tool for a man looking to domesticate the human race, but Galton completely misread his results on regression. Human heights do not have a tendency to become more average with each generation. But Galton was convinced that they did. In his mind, regression was not simply a statistical device, it was a real biological force.

There is a certain irony in the fact that the person who discovered correlation couldn't interpret his own results because he was still thinking in terms of simple cause and effect. But Galton's misunderstanding was to prove costly. His stubborn belief in the all-importance of heredity led him to overlook the influence of the environment, and head off in search of explanations for a bogus biological law. He constructed elaborate hereditary ideas, and indulged in extravagant thought experiments, dragging up the dregs of the pangenesis theory in irregular and highly indigestible pieces.

Galton brought together a collection of his hereditary thoughts in his 1889 book, *Natural Inheritance*. In one sense, the book was hugely influential. With his statistical foundations, Galton was giving heredity its first taste of theoretical respectability. But much of the book's biological content was either wrong or incomprehensible. Even worse, his literal belief in regression was forcing him to question his faith in Darwinian evolution.

Darwin believed that organisms evolve gradually, through small, incremental changes. In a statistical sense, natural selection would produce slow and subtle shifts in the average values of characteristics. But how could this happen, Galton asked, if regression was always pulling individuals back towards the population average? Natural selection might stretch a population in one direction or the other, but the moment the tension was relaxed, regression would simply return it to its original position.

With these considerations in mind, Galton felt he had no choice but to reject Darwin's idea of gradual change and invoke a new evolutionary protocol. If natural selection could never overcome the gravitational pull of regression, then the only way to evolve was by leaping to an entirely new position of stability. Small steps were out; giant jumps, Galton argued, were the only way forward for evolution.

Galton was not the first to dismiss Darwin's gradualist stance in favour of large, discontinuous steps. Thomas Henry Huxley had already pointed to the staggered series within the fossil record as evidence that evolution jumps rather than shuffles. And other big names, like the English evolutionist, William Bateson, and the Dutch botanist, Hugo de Vries, would join Galton on the growing bandwagon of biologists who subscribed to the philosophy of evolution by jerks.

Living evidence of these evolutionary leaps came, it was claimed, from the occasional birth of individuals who were unusually different from their parents and every other member of the population. In fact, the word 'mutation' was originally coined to describe the rare appearance of these so-called 'sports' of nature. Huxley pointed to the Ancon sheep, a short-legged variety, to illustrate the idea. Galton plumped for a slightly more unusual example, Johann Sebastian Bach. 'Can anybody believe the modern appearance in a family of a great musician is other than a sport?' Well, yes they could, since Bach came from a long and distinguished musical family that Galton had talked about at length in *Hereditary Genius*. Perhaps he had somehow forgotten this.

Irritating details aside, sports seemed to offer salvation to the many biologists who harboured doubts over a strict reading of Darwin's evolutionary argument. For Galton, however, solving one problem merely stirred up another. Sports were thought to be so distinct that they were considered incapable of breeding with their ancestors; Hugo de Vries considered them as founders of entirely new species. This did raise certain practical difficulties. How, for instance, would a single sport propagate a new species? Who would it mate with? De Vries got round this thorny issue by conjuring up the idea of mutating periods – rapid bursts of mutation during which numerous, reproductively compatible, sports were created simultaneously.

Galton's human evidence, however, seemed a little tenuous. Was J.S. Bach really a new kind of human? Were the Brandenburg Concertos a novel form of courtship song? Was his music simply the sound of a lonely man in search of a mate? Probably not. Even if Bach was a new species, he certainly didn't seem to have any fertility problems. He married twice and fathered twenty children.

Yet Galton had to take these issues seriously because they directly affected the terms and conditions of his original eugenic proposals. He doubted that the human race could be improved gradually, by selection on small variations in human ability. Regression stood firmly in the way of that strategy. The best hope now was to identify and locate human sports and harness their hereditary potential:

It does not seem to me by any means so certain as is commonly supposed by the scientific men of the present time, that our evolution from a brute ancestry was through a series of severally imperceptible advances. Neither does it seem by any means certain that humanity must linger for an extremely long time at or about its present unsatisfactory level. As a matter of fact, the Greek race of the classical times has surpassed in natural faculty all other races before or since, and some future race may be at least the equal of the Greek, while it is reasonable to hope that when the power of heredity and the importance of preserving invaluable 'transiliences' [sports] shall have been generally recognised, effective efforts will be made to preserve them.

Galton was a little coy when it came to elaborating on what these effective efforts might be. This was not surprising. After all, the notion of sports had transparent implications that made his eugenic ideal start to look very saucy indeed. Because of their inherent rarity, sports would necessarily become the super-studs of a new breed. A complete revision of sexual morality would be required. Forget about state-sponsored marriages and traditional family values. Monogamy would be brushed aside by the promiscuous and lucky few.

In the winter of 1886, Galton was awarded the Royal Society's Gold Medal for his statistical contributions to biology. This was more than just a recognition of his recent work on regression and correlation. The medal was a general acknowledgement of Galton's entire quantitative approach to heredity, anthropology and psychology. With his statistical inquiries he was breathing much-needed life into subjects previously unresponsive to scientific investigation.

After receiving the award, Galton gave a speech in which he outlined the course of his scientific career, carefully dissecting his past and the influences that had shaped him. Predictably, he claimed a hereditary hand had guided him towards both statistics and biology. 'On my father's side,' he explained, 'I know of many most striking, some truly comic, instances of statistical proclivity.' And when it came to biology, he was sure that 'there is a similarity between the form of the bent of my mind and that of my mother's father, Dr Erasmus Darwin. The resemblance, he continued, 'chiefly lies in a strong disposition to generalise upon every-day matters that commonly pass unnoticed'. It was through his grandfather's hereditary influence, he insisted, that he was 'well prepared to assimilate the theories of Charles Darwin when they first appeared in his *Origin of Species*'. At this point, Galton's speech took a subtle shift in tone. Memories filled the room as he lamented his lost idol:

Few can have been more profoundly influenced than I was by his publications. They enlarged the horizon of my ideas. I drew from them the breath of a fuller scientific life, and I owe more of my later scientific impulses to the influences of Charles Darwin than I can easily express. I rarely approached his genial presence without an almost overwhelming sense of devotion and reverence, and I valued his encouragement and approbation more, perhaps, than that of the whole world besides.

Accumulating numbers: Galton, aged sixty-six

Darwin was now gone for good, his death in 1882 casting a 'deep gloom' over the Galton household. But for Galton, the memory of his mentor would never fade. He kept a small miniature of Darwin on his desk at Rutland Gate. It offered a permanent and poignant reminder of the past, and the inspiration that continued to carry him forward.

The 1880s were as productive as any other period in Galton's life. The closure of the International Health Exhibition in 1885 had forced a temporary suspension of activities at his anthropo-metric laboratory. But Galton hastily organised a relocation to the Science Galleries of the South Kensington Museum so that the counting could continue. Over the next few years, many thousands more people would make themselves available for measurement. Indeed, it was a measure of how far Galton's celebrity had grown when the Prime Minister, William Gladstone, dropped by for a visit:

> Mr. Gladstone was amusingly insistent about the size of his head, saying that hatters often told him that he had an Aberdeenshire head – 'a fact which you may be sure I do not forget to tell my Scotch constituents'. It was a beautifully shaped head, though low, but after all it was not so very large in circumference.

While awards and accolades provided welcome recognition of his achievements, they could do nothing to conceal his advancing years. Galton was now well into his sixties and his failing health was an increasing concern. Throughout the 1880s recurrent bouts of gout, gastric fever and bronchitis undermined the prodigious work rate he had established for himself. Louisa, too, was regularly prone to all sorts of complaints that had her pining for the warmer and more therapeutic climes of the continent.

A serious outbreak of cholera in France and Italy kept the Galtons out of Europe in 1884 and 1885. But by 1886 the disease had died down, allowing them to recommence their continental tours. In the spring they left for Italy, heading first to Rome, where Louisa, for once, seemed invigorated. 'We were so happy there,' she wrote in her 'Annual Record', 'every passing hour had its tale of keen enjoyment and we felt to grudge each day as it passed. We returned by Amalfi and Florence where we spent another week to be remembered, such beauty and loveliness in art and nature.' But

Louisa's physical renaissance coincided with a decline in her husband's health. On their way back to Britain the couple were forced to make an extended stop at the French spa town of Contrexéville, to allow Galton to sit out a serious attack of gout.

This was, by now, a familiar pattern. Galton and Louisa never seemed to be ill at the same time. It was as if their ailments oscillated in perfect asynchrony. Louisa's condition tended to be at its worst when her husband was at his healthiest and busiest. When Galton suspended his work to tend to his wife there would usually be an immediate improvement in her health. But as Louisa started on her upward slope of recovery, she would often pass her husband on the way down. Louisa spelt out this remarkable seesaw cycle for herself in her 'Annual Record' entry for 1888: 'Frank was ill all the early part, congestion of the lung and I felt very anxious about him and overtired myself, as when he recovered, I fell ill and that brought on my old pain, which never left me until we had recourse to Vichy in [September]'.

Illness or not, Galton's capacity for counting was undiminished. During a stay in the French town of Vichy, nothing seemed too trivial to document. Out walking the streets one day, he put each woman that passed him into one of six size categories, varying from 'thin' at one extreme to 'prize fat' at the other. The observations continued over two consecutive days, Galton recording five times more prize fatties on day two than on day one. On another occasion he measured out a distance of seven and a quarter yards between two landmarks on the street, and then retreated to an inconspicuous vantage point. As various people walked by he recorded the number of paces and the time taken to walk the seven and a quarter yards. A 'youth', for instance, walked nine and a half paces in four and three-quarter seconds. A 'gentlemen' covered the distance in nine paces and five seconds, while a 'man in black coat' took only eight paces in the same amount of time.

While these episodes of continental counting were undoubtedly illuminating, it was back in Britain where Galton pulled off his real coup. Throughout the 1880s Victorian citizens may have been perturbed by the sight of a bald, elderly man, with his hand in his pocket, weighing up women on the street. It may have looked like lechery, but Galton preferred to see it as another solid statistical survey. His unusual behaviour was all part of his effort to create a beauty map of Great Britain, to rate and rank its cities on the basis of their feminine allure. As he was extremely anxious about attracting attention, he thought that he would look less conspicuous if he utilised a specially made recording device that could be concealed in his pocket.

> Whenever I have occasion to classify the persons I meet into three classes, 'good, medium, bad', I use a needle mounted as a pricker, wherewith to prick holes, unseen, in a piece of paper, torn rudely into a cross with a long leg. I use its upper end for 'good', the cross-arm for 'medium', the lower end for 'bad'. The prick-holes keep distinct, and are easily read off at leisure. The object, place, and date are written on the paper. I used this plan for my beauty data, classifying the girls I passed in streets or elsewhere as attractive, indifferent, or repellent.

Galton spent months fiddling with his pricker. He admitted that his assessment was a 'purely individual estimate', but declared his result consistent, 'judging from the conformity of different attempts in the same population'. His conclusion, however, probably did little to please the Scots. 'I found London to rank highest for beauty; Aberdeen lowest,' he pronounced.

Aberdeen's women may have felt aggrieved at the title of ugliest in Britain, but at least their honesty was never called into question. The town of Salonika, in Greece, was not so lucky. The details of Galton's second survey are vague, but from his extensive field research he somehow came to the conclusion that Salonika was Europe's capital for lying.

Galton's accumulation of numbers never stopped. Even in the most unlikely of circumstances, the counting could find an outlet.

During an especially dull meeting of the Royal Geographical Society, he dreamed up a boredom index, based on the frequency of fidgeting in an audience. Focusing his attention on just a small section of those assembled, he began counting the total number of fidgets per minute. Fearful that the use of a watch might attract attention, he measured time by counting his breaths, of which he knew there were exactly fifteen in a minute. Measurements had to be made when the audience was both attentive and indifferent, so as to estimate what Galton called the level of 'natural fidget'. Armed with this average he reckoned that the relative boredom of any audience could be gauged. He was careful, however, to add the following important proviso: 'These observations should be confined to persons of middle age. Children are rarely still, while elderly philosophers will sometimes remain rigid for minutes.'

14

Tips on Fingers

Some mechanism ought to be devised for shaking elderly people in a healthful way, and in many directions.

Francis Galton

In the early 1890s the orbit of Mars brought it unusually close to Earth, and there was a great deal of speculation in the Victorian press about extraterrestrial communication and the existence of Martians. Although many commentators accepted that it might – just – be possible to send signals to the red planet, it was generally agreed that any kind of signalling system could do little more than establish the existence of intelligent life on Mars. A serious conversation with the Martians seemed out of the question.

Galton, however, was far more optimistic about the possibility of an inter-planetary exchange, and he set about devising a complex signalling system that, he believed, would form the rudiments of a celestial language. Assuming that any Martians on the other end of the line would be fully versed in arithmetic and mathematics, he built a step-by-step code that would enable the Martians to send mathematical descriptions of themselves and their society.

While he waited for a Martian reply, Galton indulged in a whimsical fantasy about his potential alien acquaintances. He imagined a Martian community composed of fertile females, neuter females and fertile males. Even on Mars, it seemed, Galton could not escape from his eugenic obsession. It was a case of new planet, same society:

> The fertile females queen it over the males. They are far superior to any of them in size and strength and such is the constitution of the sex that while their figures and imposing demeanour excite some fear in the weaker males they also invoke their chivalrous loyalty and attachment. The fertile females rarely associate; they do not necessarily dislike one another but they are too jealous and self-contained for friendship.
>
> The neuter females possess no quality that we should call loveable: they are devoid of generosity and they have but little originality, but they continually occupy themselves with work of some sort however petty and cannot keep still for a moment. Whatever passion they possess is socialistic – they certainly care little for themselves and much for the community, and though very obstinate in small things are practically directed by the males with the concurrence of the queens, towards whom their attitude of mind is peculiar. It is one neither of personal love nor of personal loyalty but rather one of respect for the temporary representative of a necessary institution which prevents the community from becoming extinct.
>
> The males are warriors. They have all the truly male virtues and all the male defects of our race; they are the salt of the community, both morally and intellectually. They look on the neuters as 'hands' in a factory, not particularly disliking them, but as members of a different social stratum, who have to be dealt with in matters of business but not as friends.

Back on earth, Galton was consolidating his work at the anthropometric laboratory in South Kensington. The laboratory had become something of a local London attraction, although its popularity had little to do with public or scientific support for

ANTHROPOMETRIC
LABORATORY
For the measurement in various
ways of Human Form and Faculty.

Entered from the Science Collection of the S. Kensington Museum.

This laboratory is established by Mr. Francis Galton for
the following purposes:—

1. For the use of those who desire to be accurately measured in many ways, either to obtain timely
warning of remediable faults in development, or to
learn their powers.

2. For keeping a methodical register of the principal measurements of each person, of which he
may at any future time obtain a copy under reasonable restrictions. His initials and date of birth will
be entered in the register, but not his name. The
names are indexed in a separate book.

3. For supplying information on the methods,
practice, and uses of human measurement.

4. For anthropometric experiment and research,
and for obtaining data for statistical discussion.

Charges for making the principal measurements:
THREEPENCE each, to those who are already on the Register.
FOURPENCE each, to those who are not:— one page of the
Register will thenceforward be assigned to them, and a few extra
measurements will be made, chiefly for future identification.

The Superintendent is charged with the control of the laboratory
and with determining in each case, which, if any, of the extra measurements may be made, and under what conditions.

H. & W. Brown, Printers, 20 Fulham Road, S.W.

Advertisement for Galton's anthropometric laboratory in South Kensington eugenics. Most of those passing through its doors had no interest in Galton's grand social ambitions. The public flocked to his laboratory for its novelty value rather than any higher scientific purpose. For all his efforts, Galton was little more than a ringmaster at a very unusual kind of circus.

Galton's anthropological work offered the first real glimpse of the physical and psychological variety endemic to the British population, or at least that portion of it that could be bothered to fill in one of his questionnaires or pay his laboratory's entrance fee. As a snapshot of a specific time and place in human history, it was, and still is, a treasure chest of statistical information. But what did it all mean in terms of his eugenic cause? In 1869 Galton had declared eminence as the universal indicator of the ideal breed. But it was a hollow and unsatisfactory assertion. Acknowledging the ambiguity of his definition, he had set off in search of more objective criteria. In one sense, Galton's subsequent psychological and anthropological work can be seen as prospecting for new possibilities. He was looking for the birthmark of genius, a vital link between the physical and the psychological, a biological bar code to distinguish the good breeding stock from the bad.

Galton's decision to add fingerprints to his shopping list of human measurements caught a wave of public interest in new methods of criminal identification. Victorian policing had been, by and large, an entirely clueless exercise. Apart from the occasional footprint or bullet left at the scene of a crime, the police relied, almost exclusively, on eye-witness accounts to bring criminals to justice. Once a criminal was caught, the more enlightened officers might take photographs of distinguishing bodily features, like moles or tattoos, so that the criminal could be identified if they offended again. But these measures had only limited effect. More often than not, repeat offenders could avoid detection simply by changing their names.

The inadequacies of the policing system were exposed by the infamous and interminable Tichborne trial of the 1870s. The trial itself centred around the fate of Roger Tichborne, the son of wealthy English aristocrats, who, having become estranged from his family over a love affair, had emigrated to South America.

In 1854 Tichborne had taken a boat from Rio de Janeiro bound for Jamaica. When the ship didn't make it to its destination it was assumed that it had sunk, and that Tichborne, along with all the other passengers and crew, had drowned. But Lady Tichborne, Roger's mother, refused to accept her son's death. Clinging to the hope that he may have been rescued, she sent advertisements around the world appealing for information.

For ten years she heard nothing. But in 1865 she received a letter from a man in Australia claiming to be her son. Arthur Orton looked nothing like the deceased Roger Tichborne, but Orton believed that Lady Tichborne's longing was so great that she might be deceived. His assumption proved correct. When the two met in Paris for the first time, she accepted him as her long-lost son. Lady Tichborne granted Orton a £1,000-a-year allowance and welcomed him, his wife, and children into her home.

Orton had already done some research into the Tichbornes, courtesy of one of the family's ex-employees. But his knowledge of the real Roger was riddled with holes, and every member of the family, bar Lady Tichborne, saw right through the scam. Still, Orton pushed on with his audacious fraud. He had his eyes on the Tichborne family fortune. And to get at that, he would have to prove that he was Roger Tichborne in front of a judge and jury.

During the trial over one hundred people testified that Orton was the real Roger Tichborne. But when Orton himself took the witness stand there was no hiding the truth. Orton couldn't provide any details of the secret engagement that had led Roger to desert his family in the first place, and his testimony collapsed under cross-examination. Not only did Orton lose his suit, he was now faced with a serious charge of perjury.

Orton wasn't going to go down without a fight, and the second trial turned out to be twice as long as the first. Orton had, once again, constructed an elaborate story with witnesses to back him up. He argued that he had been one of the few survivors of the original sinking, and had been picked up by another boat, the *Osprey*, en route to Australia. An alleged crew member of the *Osprey* took the witness stand to support the story. But the testimony fell apart when prosecutors exposed the *Osprey* as a fictitious ship and the supposed sailor as an ex-convict just out of jail. Orton was found guilty of perjury and sentenced to fourteen years in prison.

The trial was an immensely time-consuming, tedious, and expensive exercise. Everyone within the police and judiciary agreed that something had to be done to avoid a repeat performance. A method was needed that could quickly and easily prove the identity of an individual, and attention turned to France, where a new system was already being put through its paces.

In Paris, a young police clerk called Alphonse Bertillon had developed an identification system based on the principle that every person's physique is unique. Two people may share the same height, but it is extremely unlikely that they would also share the same-sized arms, feet, head, and other bodily features. By taking about a dozen measurements, Bertillon could build up a unique physical profile of any individual.

Physical profiling was only one part of Bertillon's system. His real innovation was to come up with a simple and fast indexing system, so that a set of measurements could be compared against existing records. After all, how do you go about finding a possible match among a register than contains the records of hundreds of thousands of people? Bertillon's solution was to assign each of the body measurements into one of three size categories: small, medium, and large. By storing and arranging the records in accordance with these divisions, finding a matching file translated into a series of three-way splits. Each measurement was like a junction that progressively narrowed the search, until a cabinet containing a manageable number of individual files was located.

Galton was far more interested in the identification of genius than the detection of criminals. But for a man who still believed that the psychological could manifest itself in the physical, the two quests were almost one and the same. In his hunt for the mark of

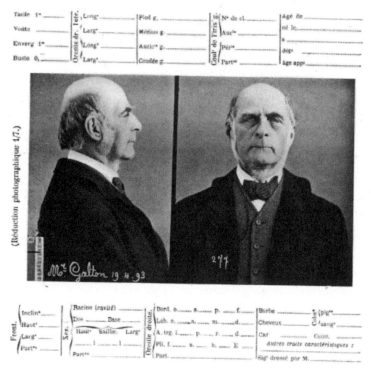

Mug-shots of Galton taken during his visit to Bertillion's Paris laboratory, 1893

genius, Galton had visited Bertillon's police department in Paris. The Frenchman's system of physical profiling offered Galton little that he had not tried for himself. Fingerprints, on the other hand, offered a real alternative:

> My attention was first drawn to the [fingerprint] ridges in 1888, when preparing a lecture on Personal Identification for the Royal Institution, which had for its principal object an account of the anthropometric method of Bertillon, then newly introduced into the prison administration of France. Wishing to treat of the subject generally, and having a vague knowledge of the value sometimes assigned to finger marks, I made inquiries, and was surprised to find, both how much had been done, and how much there remained to do, before establishing their theoretical value and practical utility.

William Herschel, a civil servant and amateur scientist, had already gained some experience in the use of fingerprints during his twenty-year stint as a chief administrator in India. Stationed to a rural province of Bengal, he had been left in charge of a population deeply resentful of British rule. Herschel had to rely on local people to get things done, but written contracts meant nothing when the signatures that were supposed to bind them were routinely disowned.

Amid this atmosphere of civil disobedience Herschel hit upon a new system of identity that, he believed, might stifle any contractual deception. In 1859 he made an arrangement with Raj Konai, a local building contractor, for the supply of road-building materials. After the contract was drawn up Herschel turned the paper face down on his desk. Holding Konai's right hand, he smeared it in ink before pressing it down onto the contract. His hand left a perfect impression on the paper, complete with the detailed lines and ridges of his palm and fingers. A man may deny his signature, but Herschel felt sure that he could never disown the imprint of his hand. His hunch was correct and Konai honoured the contract.

The use of handprints or fingerprints as statements of identity did not originate with Herschel. The Chinese and Japanese had been employing them for centuries, perhaps from as early as AD 600. In fact Herschel may have come across their use in Calcutta, where there was a large Chinese population. Wherever he got the idea, he was keen to take things further and capitalise on his initial success. Following the Chinese lead he switched from the awkward and messy palm print to the more manageable fingerprint.

Over the next twenty years Herschel led his own personal investigation into the fingerprint phenomenon, taking literally thousands of prints from anyone who was prepared to get their fingers dirty. Convinced that an individual's prints were both unique and permanent, he urged the Indian authorities to adopt the use of fingerprints on all kinds of legal documents. Officialdom, however, did not share his enthusiasm and it wasn't until 1877, when Herschel had risen to the seniority of a governing magistrate, that he could personally authorise their introduction for himself.

Herschel may have considered himself the number-one fan of the fingerprint, but he had a rival in the form of Henry Faulds, a Scottish medical missionary, who had conducted his own study of fingerprints in Japan. When Faulds returned to Britain in the mid-1880s, he spent two years trying to convince Scotland Yard that fingerprinting was the answer to every policeman's prayers. Faulds wasn't just trying to flog a general idea. He claimed to have the whole system worked out, from the way that the prints should be taken, to the way they could be classified and indexed. But the police were not convinced. Despite his exemplary sales pitch, Scotland Yard turned away from Faulds and his fingerprints. Scientifically speaking, he was an unknown quantity, a maverick with unproven credentials.

Galton was familiar with the work of both Herschel and Faulds, but it was Herschel to whom he first turned for assistance. Herschel was the grandson of his more famous namesake, the eminent astronomer who discovered the planet Uranus. The whole family had a rich scientific heritage and had featured prominently in Galton's *Hereditary Genius*.

With Herschel signed up to provide help, Galton outlined the task ahead of him:

> Three facts had to be established before it would be possible to advocate the use of finger-prints for criminal or other investigations. First, it must be proved, not assumed, that the pattern of a finger-print is constant throughout life. Secondly, that the variety of patterns is really very great. Thirdly, that they admit of being so classified, or 'lexiconised', that when a set of them is submitted to an expert, it would be possible for him to tell, by reference to a suitable dictionary, or its equivalent, whether a similar set had been already registered. These things I did, but they required much labour.

Galton spent much of the late 1880s and early 1890s scrutinising his vast stockpile of fingerprints. His first priority was to work out a system of classification. Only then would he be able to make the racial or intellectual distinctions he was hoping to find. In defining fingerprint patterns in terms of loops, arches, and whorls, Galton followed Henry Faulds's system of classification. But Galton also looked beyond these more obvious differences, going deep into the minutiae of the prints themselves.

A fingerprint is an impression of the tiny ridges that project from the surface of the skin. For much of their route across the finger tip these ridges run a tight and parallel path. But every now and again a ridge might form a junction with its neighbour, like a miniature set of points on a train track. Sometimes a track splits in two, only to rejoin further down the line. Some tracks come to abrupt dead ends, while others have sidings that lead nowhere. As Galton discovered, no two fingers ever share the same system of tracks. It was as if there was a unique rail network sitting on the end of each finger.

Having identified these miniature marks of distinction, Galton proceeded to prove their permanence. With William Herschel's help, he managed to obtain samples of fingerprints taken twenty years apart. When he studied them carefully he found that the prints were identical in every detail. Nothing had changed. Even if the ridges were destroyed by a cut or a burn they would simply grow back in their original form. Only a deep wound could threaten their complete revival.

While the minutiae provided many detailed insights, Galton found that he couldn't easily incorporate them into a simple index. So, like Bertillon before him, he decided to build his system of filing and retrieval on superficial distinctions. Instead of small, medium, and large, each digit was classified as A (arch), L (loop), or W (whorl), so that a complete set of prints was indexed by a sequence of ten letters. The sequence read like a series of directions, guiding the investigator ever closer to his destination. While it could never pinpoint the exact location, it could bring him to within a few files of home.

That fingerprint patterns fell neatly into the three distinct categories of arch, loop, and whorl hit a chord with Galton. His years of studying fingerprints were during a period when he was preoccupied with the idea of regression as a restraining force in evolution. In his mind, natural selection could never make any long-term evolutionary progress so long as it was working on minor variations, because regression was always pulling a population back towards the average. The only way to evolve was by giant leaps to entirely new positions of stability. Fingerprints, he believed, fitted in perfectly with this perverted picture of evolution. He saw arches, loops and whorls as living examples of evolutionary jumps, stopping-off points in the hopscotch path of evolution.

Fingerprints may have affirmed his erroneous views on evolution, but they were not the trump card he'd been hoping for. When Galton looked at samples of fingerprints collected from English, Welsh, Jewish, and African schoolchildren he could draw no obvious distinctions between them. Clearly, it was not the result he'd anticipated, and he sought solace in a hunch:

> Still, whether it be from pure fancy on my part, or from some real peculiarity, the general aspect of the Negro print strikes me as characteristic. The width of the ridges seems more uniform, their intervals more regular, and their courses more parallel than with us. In short, they give an idea of greater simplicity, due to causes that I have not yet succeeded in submitting to the test of measurement.

Race was not the only distinction to be squashed under the thumb. Eminence, too, forged no mark on the finger. Galton wrote, 'I have prints of eminent thinkers, and of eminent statesmen that can be matched by those of congenital idiots. No indications of temperament, character, or ability can be found in finger marks, so far as I have been able to discover.'

He was down, but not defeated. Fingerprints may have been worthless as a eugenic tool, but they could still be put to good use in criminal investigations, in a national system of identification, and 'in our tropical settlements, where the individual members of the swarms of dark and yellow-skinned races are mostly unable to sign their names and are otherwise hardly distinguishable by Europeans, and, whether they can write or not, are grossly addicted to personation and other varieties of fraudulent practice'.

Galton summarised his work on fingerprints in an eponymously titled book, published in 1892. While *Finger Prints* holds the honour of being the first systematic study of the subject, it was Herschel and Faulds, of course, who were the original pioneers. Galton had simply turned their speculative ideas into something much more substantial. Unfortunately, that was not the impression that readers got from reading the book. Galton wrote all about the important contributions of William Herschel, but the name of Henry Faulds was suspiciously scarce. Even when the Scottish doctor did get a mention, Galton got his name wrong.

In 1893 the Home Secretary, Herbert Asquith, appointed a select committee to investigate ways of overhauling the British system of criminal identification. The so-called Troup committee was asked to consider all the available options, including both fingerprints and 'Bertillonage'. During their investigation, the committee called in on Galton's anthropometric laboratory, to see for themselves how fingerprints could be taken, indexed, and stored. Although the committee was highly impressed by Galton's set-up, there were concerns over his indexing system. In the end, it was decided that a combination of Bertillonage and fingerprints would provide the best solution.

Despite the compromise, this was the first public endorsement of fingerprints, and Galton's name was all over the newspapers. The stories did not make pleasant reading for Henry Faulds. He had spent two years trying to impress the value of fingerprints on a sceptical Scotland Yard. Now, seven years later, the police had finally changed their mind. But it was Galton who was getting all the glory.

FINGER PRINTS

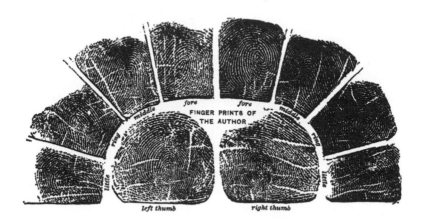

BY

FRANCIS GALTON, F.R.S., ETC.

Galton's seminal work on fingerprints, with a set of his own prints on the cover

Faulds was well within his rights to feel aggrieved about the way he was being treated. After all, it was he, and not Herschel, who had first communicated his findings to *Nature* in 1880. But Galton saw things differently. In return for the help he had received from Herschel, Galton used his own considerable scientific influence to promote Herschel as the ultimate fingerprint pioneer. Herschel more or less confessed to their gentleman's agreement years later in a letter to *The Times*:

When communicating to [Galton] in 1888 ... I merely stipulated that he would recognise the fact of my putting the finger-print system into full and effective work ... as early as 1877, after some 20 years' experimenting for this one definite purpose. He did more than keep his promise at all times . . . He assigned to me the priority of devising and adopting officially a feasible method of turning the finger-marks to practical use for identification.

From Galton's perspective, the move to promote Herschel made perfect sense. Herschel descended from a distinguished family of eminent men, whereas Faulds came from a far more modest social background. It was only natural, therefore, that Faulds should be the name that was erased from history.

Faulds became so incensed by the snub that he challenged Galton and Herschel to a bare-knuckled fight. But Galton pre-ferred to throw punches with words rather than fists. When Faulds wrote a fingerprint book of his own, Galton greeted it with a scathing review in *Nature*:

Dr. Faulds in his present volume recapitulates his old grievance with no less bitterness than formerly. He overstates the value of his own work, belittles that of others, and carps at evidence recently given in criminal cases. His book is not only biased and imperfect, but unfortunately it contains nothing new that is of value.

By the late 1890s Galton had lost his grip on the future of fingerprinting, and it fell to Edward Henry to develop a more user-friendly and comprehensive system of fingerprint classifica-tion. Henry had been working as Inspector-General of Police on William Herschel's old patch in Bengal, where fingerprints were still going strong. Appointed Chief Commissioner of the Metro-politan Police in 1901, Henry introduced a criminal-identification system that was based exclusively on fingerprints. His Fingerprint Branch at New Scotland Yard set new forensic standards that were adopted by police forces around the world.

The creation of the Fingerprint Branch was symptomatic of the wider changes taking place within British society. As the century neared its end, a more sophisticated world was emerging. Radical reforms in education, health, and voting rights were matched by technological revolutions in communications, transport, and medicine. Antiseptics and anaesthetics, telephones, trams, electric lighting, and the internal combustion engine all heralded the arrival of a more modern age. The building of grand civic universities in Birmingham, Manchester, and other industrial cities reflected both the rapid growth in science, and its increasingly professional outlook. In this climate of change, Galton's own brand of scientific endeavour was starting to look quaint and old-fashioned. The era of the polymath was almost over.

Galton himself seemed to be in serious decline. His increasing deafness was now keeping him away from all but the most important scientific meetings, while recurring bouts of asthma continued to eat into his schedule, slowing his work rate to a fraction of its former pace. 1895 and 1896 turned out to be two of his least productive years in decades, his output reduced to occasional correspondence with newspapers or magazines. One letter, submitted to the *Spectator*, told the story of a crazy cat called Phyllis, the matriarch to a whole family of demented felines. 'Three Generations of Lunatic Cats' suggested that the madness was an inherited characteristic.

Despite the deterioration in the health of both Galton and Louisa, the couple continued to take their continental tours. In the summer of 1895 they were in southern Germany, first enjoying the Alpine scenery of Garmisch, before heading north to the medieval sights of Rothenburg, Nuremberg, Frankfurt and Bonn. The Galtons were back in southern Germany the following year, this time for a therapeutic cure at the hot baths of Wildbad. Louisa outlined events in her 'Annual Record' for 1896:

> Frank fell ill on the 19th June with Gastric Catarrh like he had before, but much milder and it might have been nothing, had he not gone to Kew with Temperature 102 so with guests

and nursing I was nearly done for. [Dr] Chepmell advised
Wildbad as a cure and tho' still very weak but in perfect
summer weather, he and I left England July 10 and began the
treatment July 14.

The waters did wonders for Galton's wheezing, and enabled the
couple to enjoy another long tour of Germany and Switzerland.
But this was to be their last full summer together. In 1897 Louisa
made no entry to her 'Annual Record'. Instead, it was Galton who
took up the story. 'It is with painful reluctance that I set down the
incidents of this fatal year, and do so on Jan. 6, the anniversary of
the day when I first became acquainted with dear Louisa at the
Dean's house, next door to our own, at Dover in 1853.'

In August 1897 the couple were on holiday in central France,
when Louisa fell ill with diarrhoea and nausea. At first there was
nothing to indicate that this was anything more than a recurrence
of a common complaint. But her condition deteriorated rapidly
and she died a few days later. Writing from his hotel room at Royat
in France, Galton explained all in a letter to his eighty-six-year-old
sister:

Dearest Emma, It is ill news that I have to send. You heard
that Louisa had been ill since last Sunday, when she packed
up in good spirits and with much interest for a tour among
the Dauphiné mountains, beginning with the Grand Char-
treuse. But it was not to be. She was seized with a severe
attack of diarrhoea and vomiting during the night, a repeti-
tion of what she, I and Mme de Falbe had all had in a lesser
degree. Still, Dr Petit thought little of it on Monday morning,
even on Tuesday morning he was not anxious, but she
grew steadily worse. The bile thrown out was exceedingly
disordered and I think its presence throughout the body
poisoned her. She had of course discomfort at times, but
was on the whole drowsy. Yesterday she was evidently sinking. I had a nurse to sit up through the night, who awoke me

at 2½ a.m. when dear Louie was dying. She passed away so imperceptibly that I could not tell when, within several minutes. Dying is often easy! . . . I cannot yet realise my loss. The sense of it will come only too distressfully soon, when I reach my desolate home . . . Dear Louisa, she lies looking peaceful but worn, in the next room to where I am writing, with a door between. I have much to be thankful for in having had her society and love for so long. I know how you loved her and will sympathise with me. God bless you. Ever affectionately, Francis Galton.

A postmortem examination revealed that Louisa's stomach had shrunk to one-third of its natural size, while the exit tube leading to her duodenum was barely the width of a pencil. With such a severe internal constriction – possibly triggered by her serious illness in Dorset nineteen years earlier – it was little wonder that she had suffered from perennial ill health.

Galton, shaken but philosophical, organised her burial in the cemetery at Clermont-Ferrand and then returned to Britain, seeking sympathy from friends and relatives. For a while it looked as though fingerprints would be his final say on science. But as the weeks passed, he gradually began to emerge from his torpor. Louisa's death seemed to have issued a wake-up call. With his wife gone, Galton had lost the counterpoint to his own oscillating illnesses. He was suddenly free again, invigorated, and raring to go, ready to make his last giant leap into history.

15

Home Improvements

Whoever struggles with monsters might watch that he does not thereby become a monster.

Nietzsche

Galton spent much of his life arguing that people were complex but measurable machines. Given the right perspective, any aspect of mind and body could be reduced to a numerical from. The loss of Louisa must have seemed like the ideal opportunity to put his principles into practice and quantify his own grief. He seemed genuinely affected by the death of his wife, but if he ever did measure his feelings, he never made them public. Instead, he tried to put the pain of others into a numerical context. In 1898 *The Times* conducted a debate on corporal punishment. Galton's contribution was a letter outlining two important points that he felt had been overlooked:

> The first is, the worse the criminal the less sensitive he is to pain, the correlation between the bluntness of the moral feelings and those of the bodily sensations being very marked. The second relates to the connection between the force of the blow and the pain it occasions, which do not vary at the same rate, but approximately, according to Weber's law, four times as heavy a blow only producing about twice as much pain. In

a Utopia the business of the Judge would be confined to sentencing the criminal to so many units of pain in such and such a form, leaving it to anthropologists skilled in that branch of their science to make preliminary experiments and to work out tables to determine the amount of whipping or whatever it might be that would produce the desired results.

When it came to capital punishment, units of pain were of little consequence, but Galton still found some grave issues to consider. Death by hanging was the common mode of execution in Britain. The 'long drop' was intended to break the neck, the length of fall being based on the height and weight of the victim. But on one occasion, the distance was patently overestimated when an exceptionally large man lost not only his life, but also his head. It was Galton who spotted an error in the formula. He communicated his findings to *Nature*.

Morbid thoughts were evidently playing on Galton's mind in the wake of his wife's demise. In a letter to his sister Emma he had sounded gloomily fatalistic. 'I have very much to be grateful for, but our long-continued wedded life must anyhow have come to an end before long. We have had our day, but I did not expect to be the survivor.' For once, his hereditary devotion had deserted him. The fact that most of his elder brothers and sisters were still alive was surely evidence enough that Galton had years left in him. Lucy, admittedly, had died young, and Adèle was seventy-three when she passed away in 1883. But at the time of Louisa's death in 1897, Bessy, aged eighty-nine, Emma, eighty-six, Darwin, eighty-three, and Erasmus, eighty-two, were all going strong. At seventy-five, Galton was still the youngster in the family.

He rediscovered some of his old vigour when a new woman walked into his life. His family had suggested that Eva Biggs, granddaughter of his sister Lucy, might become his personal nurse and companion. Eva was put through her paces on a long tour of southern Spain, beginning in March 1899. She and Galton took a boat to Gibraltar before travelling inland to explore the Andalucian cities of Granada, Malaga, and Seville. The two of them hit it off immediately. Rarely had Galton sounded so effusive about a member of the opposite sex. 'Eva is a capital companion,' he wrote to his sister, Emma, 'always cheerful and punctual and interested; moreover she always sees the good sides of things and of persons.' Despite the age gap – Eva was forty years his junior – they seemed perfectly suited to one another. From Granada, Eva wrote to her aunt Emma:

> I don't suppose he ever mentions his cough, so I will tell you about it. It has never actually gone yet, but is much better and he looks very well and is tremendously energetic, the Spaniards all ask me his age, and won't believe it when I tell them; you should see his complexion when he is on the sea, it is splendid . . . He really is a perfect person to travel with, because he never fusses or gets impatient *or* grumbles if we are kept waiting ever so long for food or luggage! . . . I am so happy out here, I love the Spaniards, they are so kind and polite to us, but all the same they are poor creatures and not a bit strong-minded or intellectual, but so picturesque. We hardly ever see any English, if we do they are men and rather second-rate. Plenty of fat overfed Frenchmen!

The Spanish tour set the pattern for the next few years of Galton's life. From now on winters would be spent abroad, in a climate more sympathetic to his weak chest. Rutland Gate remained his London home, but he cut his scientific commitments to a minimum, allowing himself more time and freedom to travel. In September 1899 he attended the annual meeting of the British Association in Dover. But come the autumn, he and Eva were abroad again, this time in Egypt.

It was a nostalgic trip, and a sad one. Galton had not been to Egypt since his tour of the 1840s. He still had fond memories of the French explorer, d'Arnaud Bey, the man who had given direction

Evelyne Biggs, Galton's nurse and companion in his last decade

to the aimless wanderings of his twenties. In an effort to track him down, Galton called in on Egypt's Geographical Society and 'learnt how important and honoured a place d'Arnaud Bey had occupied in its history'.

> He had died not many months previously, and I looked at his portrait with regret and kindly remembrance. Being asked to communicate a brief memoir to the Society at its approaching meeting, I selected for my subject a comparison between Egypt then and fifty years previously. I took that opportunity to express my heartfelt gratitude to d'Arnaud, which posthumous tribute was all I had the power to pay.

Egypt was a very different country to the one that Galton had visited as a young man. The Suez Canal had turned it into a strategic stronghold with a large British presence. Military men walked the streets of Cairo, wealthy tourists steamed down the slowly shrinking Nile, and scientists swarmed all over its archaeological treasures.

Galton's Egyptian tour included a week-long stay at Abydos, on the western side of the Nile between Thebes and Cairo, where the British Egyptologist, Sir Flinders Petrie, had established an archaeological dig. Accommodation at the camp was basic, consisting of little more than mud huts, nine feet long by seven feet wide, with a hanging mat for a door. 'Our host and hostess were peculiarly independent of ordinary comfort,' remembered Galton, 'but the consumption of marmalade at their table was enormous.'

Despite the occasional discomforts, Galton was in good spirits. When he wasn't catching up with old scientific and society friends, he and Eva were taking in the surfeit of sights on offer. His mind, however, was distracted by events elsewhere. 'I feel very painfully', Galton confessed to his sisters, 'the contrast between my enjoying myself lazily in this glorious climate and the sufferings of our countrymen at the Cape.' Three thousand miles away, on the opposite side of the continent, the British Army was suffering heavy losses against the Boers. The army had arrived in South Africa expecting another colonial walkover. But the Boers, fortified by German guns, were making the British look disorganised and hopelessly ill-prepared.

The Boer war was a turning point in British history, and helped to crystallise national anxieties about Britain's imperial future. The country was still a colonial force to be reckoned with, but its assumed role of imperial overlord could no longer be taken for granted. France, Germany, Russia, and America were all challenging Britain for economic supremacy. Militarily, too, Britain was looking over its shoulder at the growing imperial navies of Germany, Japan, and France.

If Britain's imperial influence was on the wane, then people demanded to know why. To some observers, the country's problems were symptomatic of an underlying physical decline. In Manchester, for instance, over 70 per cent of those who had volunteered for the Boer war had been rejected on the grounds that they were physically unfit to fight. These were undoubtedly alarming statistics. But were they evidence of an innate deterioration, or was something else to blame?

Britain experienced some drastic demographic changes during the nineteenth century, as the nation's economic momentum shifted from the countryside to the towns. In the latter half of the century urban growth accelerated at an unprecedented rate. By 1901, eighty per cent of the British population lived in towns and cities, the highest proportion of any country in Europe.

City living, with its attendant social problems of crime and poverty, had always been a cause for concern. But new statistical surveys such as Charles Booth's 1901 study, *The Life and Labour of the People in London*, were now highlighting the true magnitude of the problem. Thirty per cent of the population was living in poverty, and infant mortality was at fifteen per hundred births. Although minds may have been divided on the root cause of this social malaise, there was widespread agreement that something needed to be done.

In October 1901 Galton tried to tap into these national insecurities when he gave the Huxley Lecture at the Royal Anthropological Institute. Clearly, he saw the occasion as the ideal opportunity to re-launch his eugenic initiative. 'To no nation', he declared, 'is a high human breed more necessary than to our own, for we plant our stock all over the world and lay the foundation of the dispositions and capacities of future millions of the human race.' Galton's rallying cry seemed well timed, yet once again his radical ideas attracted little publicity.

But things were beginning to change. Slowly and quietly, a small but significant band of supporters was growing. Karl Pearson, for instance, a mathematics professor at University College London, had become one of Galton's biggest fans. Won over by the austere charms of statistics after reading Galton's *Natural Inheritance*, he had established a School of Biometry at University College, dedicated to the statistical treatment of biological problems. Pearson took many of Galton's inchoate ideas and gave them a more rigorous mathematical foundation. Like Galton, he was a statistical pioneer. He was also an ardent eugenicist.

Elsewhere, events were turning more and more in Galton's favour. The belief that science might offer a credible solution to the problem of national improvement was given a significant boost by the rediscovery of Gregor Mendel's work on inheritance. Mendel's breeding studies on pea plants were originally published in the mid-1860s, but Mendel had so perfected his role as the anonymous abbot that nobody had taken much notice at the time. For thirty-five years his work remained unread and unloved. But in 1900 he was rescued from obscurity and crowned, posthumously, as the patron saint of genetics.

Unlike Galton, Mendel had not concerned himself with issues of nature and nurture. Nevertheless, the simplicity of Mendel's scheme was hugely appealing to those already sympathetic to hereditary explanations of human behaviour. If the shape and colour of a pea could be directed by pairs of hereditary particles, then perhaps the shape and colour of the human mind could be moulded in the same way. To that end, the second coming of Mendel was great publicity for Galton's eugenics because it shot heredity and, more pertinently, hereditary determinism, to the forefront of scientific consciousness.

Galton's own contribution to the study of heredity had not been forgotten. In 1902 the Royal Society awarded him the Darwin Medal in honour of his achievements. This was no endorsement of eugenics, but rather an acknowledgement of a true scientific pioneer. In his announcement of the award, Sir William Huggins offered a shrewd and honest appraisal of Galton's legacy:

> The work of Mr Galton has long occupied a unique position in evolutionary studies. His treatise on *Hereditary Genius* (1869) was not only what it claimed to be – the first attempt to investigate the special subject of the inheritance of human faculty in a statistical manner and to arrive at numerical results – but in it exact methods were for the first time applied to the general problem of heredity on a comprehensive scale. It may safely be declared that no one living had contributed more definitely to the progress of evolutionary study, whether by actual discovery or by the fruitful direction of thought, than Mr Galton.

Galton was on holiday in France when he first heard the news. A few days later word arrived of another award that gave him even greater pleasure. 'I was sure that you and Bessy and Erasmus would all be glad to hear of the Darwin medal,' he wrote to his sister Emma:

> But there is even more to tell, of even yet more value to myself. They have elected me Honorary Fellow of Trinity College, Cambridge, which is a rare distinction for a man who has not been previously an ordinary fellow, or who is not a professor resident in Cambridge . . . Is it not pleasant? This is the sort of recognition I value *most* highly. All the more so, as I

did so little *academically* at Cambridge, in large part owing to ill health. But I seem to owe almost everything to Cambridge. The high tone of thought, the thoroughness of its work, and the very high level of ability, gave me an ideal which I have never lost.

A month later, in Rome, Galton was in more sombre mood when a letter arrived from Emma, informing him of the death of his brother Darwin. 'It is more of a shock to me than I could have expected,' he wrote, 'for many happy incidents of early days crowd the memory . . . Darwin used to have a terror of death and was extremely moved if he heard unexpectedly of the death of any one he knew. Now he is initiated into the secret and has passed the veil.'

In April 1903 Galton and Eva finally returned to Rutland Gate, which had undergone some radical revisions. One especially asthmatic night in a heavily carpeted French hotel had convinced Galton that carpets did little for his coughing, and he had ordered them to be removed from his house.

In 1903 Galton produced very little new scientific work, but he did have a short piece published in the *Daily Chronicle* entitled 'Our National Physique – Prospects of the British Race – Are We Degenerating?' The article was significant, not so much for its theme, but for its tone. Galton was beginning to sound open-minded, even cautious in his advocacy of social change:

My attitude, which has usually been misrepresented, is to urge serious inquiry into specific matters which still require investigation in the well-justified hope that a material improvement in our British breed is not so Utopian an object as it may seem, but is quite feasible under the conditions just named. But whatever agencies may be brought to bear on the improvement of the British stock, whether it be in its Nature, or in its Nurture, they will be costly, and it cannot be too strongly hammered into popular recognition that a well-developed human being, capable in body and mind, is an expensive animal to rear.

In November Galton and Eva were abroad again, this time heading south to Sicily. When they returned to Britain the following April, Galton made preparations for another important lecture. He had been invited to give an address at the forthcoming meeting of the newly formed Sociological Society. There would be some famous names in the audience, and Galton knew that their presence would guarantee publicity for his ideas. To help his cause he arranged for Karl Pearson, his chief scientific supporter, to chair the meeting.

The talk, entitled 'Eugenics: Its Definition, Scope, and Aims', offered ample evidence that Galton had made some concessions to his former hard-line stance. Whether it was the advance of age, or simply reason, his social prescription seemed to have mellowed. Although he was still biased towards academic and intellectual achievement, he recognised that a functioning society needed a broad range of excellence. 'The aim of eugenics', he declared, 'is to represent each class or sect by its best specimens; that done, to leave them to work out their own civilisation in their own way.' Galton was using 'class' in the context of different professions, rather than different social castes, and he seemed to be anticipating a community that embraced the best plumbers, carpenters, and farriers, as well as his favoured statesmen, scientists, and athletes.

When it came to the characteristics to be favoured Galton was again quite specific. 'A considerable list of qualities can easily be compiled that nearly everyone except 'cranks' would take into account when picking out the best specimens of his class. It would include health, energy, ability, manliness, and courteous disposition.' Such exacting criteria sounded suspiciously like self-assessment, and must have left the women in the audience scratching their heads.

Galton saw the fledgling Sociological Society as a potential friend, and urged it to take an active part in his plans. He believed that the society could perform two essential functions. First, it could work as a scientific organisation, coordinating further research into human heredity. Second, it could act as a propaganda machine, going out into the nation to spread the word. Typically, he concluded his relatively brief address with a dose of his own propaganda, unleashing the kind of blathering sermon at which he had become so adept:

> What nature does blindly, slowly, and ruthlessly, man may do providently, quickly, and kindly. As it lies within his power, so it becomes his duty to work in that direction. The improvement of our stock seems to me one of the highest objects that we can reasonably attempt. We are ignorant of the ultimate destinies of humanity, but feel perfectly sure that it is as noble a work to raise its level, in the sense already explained, as it would be disgraceful to abase it. I see no impossibility in eugenics becoming a religious dogma among mankind, but its details must first be worked out sedulously in the study. Overzeal leading to hasty action would do harm, by holding out expectations of a near golden age, which will certainly be falsified and cause the science to be discredited. The first and main point is to secure the general intellectual acceptance of eugenics as a hopeful and most important study. Then let its principles work into the heart of the nation, which will gradually give practical effect to them in ways that we may not wholly foresee.

At the conclusion of Galton's talk the chairman, Karl Pearson, stood up to address the audience. Slim, suave, and handsome, Pearson was a peculiar individual, a ferocious intellectual who idolised Galton with near-psychotic reverence. Blessed with a brilliant mathematical mind, he was, at heart, a romantic idealist, a man so in love with all things German that he changed the spelling of his name from Carl to Karl. In many ways he and Galton were complete opposites. Pearson, half Galton's age, flirted with the fringes of the radical London scene. His social circle included people like George Bernard Shaw; Eleanor Marx, the daughter of Karl; Sidney and Beatrice Webb; Havelock Ellis; and Olive Schreiner.

Yet for all the overwrought idealism, Pearson, like Galton, seemed emotionally distant. Olive Schreiner, with whom he had a brief liaison, remarked that he reminded her of 'a lump of ice'. George Bernard Shaw came to a similar conclusion. 'You are never exercised on the human factor; and you come at last to be always looking for explanations under the furniture and up the chimney instead of within yourself.'

For all his faults, Pearson was the perfect propagandist, and as he primed the Sociological Society for debate, he could barely control himself. Just in case anyone in the audience was still unsure about where his own personal sympathies might lie, Pearson clarified the situation with the kind of gushing homage normally reserved for dictators and demigods:

Are we to make the whole doctrine of descent, of inheritance, and of selection of the fitter, part of our everyday life, of our social customs, and of conduct? It is the question of the study now, but tomorrow it will be the question of the marketplace, of morality, and of politics. If I wanted to know how to put a saddle on a camel's back without chafing him, I should go to Francis Galton; if I wanted to know how to manage the women of a treacherous African tribe, I should go to Francis Galton; if I wanted an instrument for measuring a snail, or an arc of latitude, I should appeal to Francis Galton; if I wanted advice on any mechanical, or any geographical, or any sociological problem, I should consult Francis Galton. In all these matters, and many others, I feel confident he would throw light on my difficulties, and I am firmly convinced that, with his eternal youth, his elasticity of mind, and his keen insight, he can aid us in seeking an answer to one of the most vital of our national problems: How is the next generation of Englishmen to be mentally and physically equal to the past generation which has provided us with the great Victorian statesmen, writers, and men of science – most of whom are now no more – but which has not entirely ceased to be as long as we can see Francis Galton in the flesh?

Galton was too deaf to participate in the discussion and remained silent and seated. He could only look on as a long list of contributors expressed various degrees of scepticism towards his proposals. H.G. Wells disagreed with Galton's assertion that criminals should be barred from breeding, arguing, 'Many eminent criminals appear to me to be persons superior in many respects – in intelligence, initiative, originality – to the average judge.' Although Wells endorsed Galton's general idea, he suggested that more emphasis should be placed on selection against the bad rather than promotion of the good. 'I believe that now and always the conscious selection of the best for reproduction will be impossible; that to propose it is to display a fundamental misunderstanding of what individuality implies . . . It is in the sterilisation of failures, and not in the selection of successes for breeding, that the possibility of an improvement of the human stock lies.'

The physician, Doctor Robert Hutchison, shared none of Wells's enthusiasm. Claiming, with some justification, to be more acquainted than most with the factors that affect physical health, he pointed to childhood nutrition as the cause for immediate concern. 'If you would give me a free hand in feeding, during infancy and from ten to eighteen years of age, the raw material that is being produced, I would guarantee to give you quite a satisfactory race as the result.' Hutchison spoke for many in the audience when he insisted, 'We should do more wisely in concentrating our attention on things such as those [nutrition], than in losing ourselves in a mass of scientific questions relating to heredity, about which, it must be admitted, in regard to the human race, we are still profoundly in ignorance.'

On a separate but related issue, John Robertson criticised Galton's belief that the success of eugenics was 'mainly a matter of the right adjustment of individual conduct, in a social system politically fixed'. This, he insisted, was far too simplistic. 'Nations can only very gradually change their hearts,' he argued, 'and part of the process consists in changing their houses, their clothes, their alimentation, their economic position, and their institutions as a means to the rest.'

The writer Benjamin Kidd queried Galton's insistence that there was widespread agreement on what constituted the fittest and most perfect individual. He went on to add a warning that, in light of the events of the Second World War, seems remarkably prescient:

> Statistical and actuarial methods alone in the study of individual faculty often carry us to very incomplete conclusions, if not corrected by larger and more scientific conceptions of the social good. I remember our chairman, in his earlier social essays, once depicted an ideally perfect state of society. I have a distinct recollection of my own sense of relief that my birth had occurred in the earlier ages of comparative barbarism. For Mr Pearson, I think, proposed to give the kind of people who now scribble on our railway carriages no more than a short shrift and the nearest lamppost. I hope we shall not seriously carry this spirit into eugenics. It might renew, in the name of science, tyrannies that it took long ages of social evolution to emerge from.

It was a sign of the changing times that women were also able to offer their opinions on the matter. Doctor Drysdale Vickery berated Galton for overlooking the role of women in his policy of race improvement. 'In the future', she opined, 'the question of population will, I hope, be considered very much from the feminine point of view; and if we wish to produce a well-developed race, we must treat our womankind a little better than we do at present.'

Perhaps the most wholehearted endorsement came from George Bernard Shaw who, though not present at the meeting, added his own written contribution. 'I agree with the paper,' he began, 'and go so far as to say that there is now no reasonable excuse for refusing to face the fact that nothing but a eugenic religion can save our civilisation from the fate that has overtaken all previous civilisations.'

It is worth pointing out that we never hesitate to carry out the negative side of eugenics with considerable zest, both on the scaffold and on the battlefield. We have never deliberately called a human being into existence for the sake of civilisation; but we have wiped out millions. We kill a Tibetan regardless of expense, and in defiance of our religion, to clear the way to Lhassa for the Englishman; but we take no really scientific steps to secure that the Englishman when he gets there, will be able to live up to our assumption of his superiority.

Shaw then took the subject of eugenic sex to places that Galton had never dared to go. 'What we need', he declared enthusiastically, 'is freedom for people who have never seen each other before, and never intend to see each other again, to produce children under certain definite public conditions, without loss of honour.' It was probably for the best that Galton was unable to hear Shaw's prurient words.

But Galton did get to read the commentaries and add his own impressions of the debate. For a man supposedly crying out for converts, he displayed remarkable hostility towards his audience. The opinions of most contributors were rubbished as nothing more than the utterances of ignorant men. Summing up, he could only conclude that 'if the society is to do any good work in this direction [eugenics], it must attack it in a much better way than the majority of speakers seem to have done tonight'.

In the aftermath of the Sociological Society meeting, Galton retreated from the spotlight. He took a short summer vacation to the Auvergne in France with his niece, Milly Lethbridge, daughter of his late sister, Adèle. Milly's recollection of events suggested that age may have conquered his shyness, but had made little impact on his energy:

> The heat was terrific, and I felt utterly exhausted, but seeing him perfectly brisk and full of energy in spite of his 82 years, dared not, for very shame, confess to my miserable condition. I recollect one terrible train-journey, when, smothered with dust and panting with heat, I had to bear his reproachful looks for drawing a curtain forward to ward off a little of the blazing sun in which he was revelling. He drew out a small thermometer which registered 94°, observing: 'Yes, only 94°. Are you aware that when the temperature of the air exceeds that of blood-heat, it is apt to be trying?' I could quite believe it! – By and by he asked me whether it would not be pleasant to wash our face and hands? I certainly thought so, but did not see how it was to be done. Then, with perfect simplicity and sublime disregard of appearances and of the astounded looks of the other occupants of our compartment, a very much 'got-up' Frenchman and two fashionably dressed Frenchwomen, he proceeded to twist his newspaper into the shape of a washhand-basin, produced an infinitesimally small bit of soap, and poured some water out of a medicine bottle, and we performed our ablutions – I fear I was too self-conscious to enjoy the proceeding, but it never occurred to him that he was doing anything unusual!

An enjoyable summer was marred by the death of Galton's sister Emma in August. It was, according to Milly, 'a terrible blow to him'. Since his days at medical school, Emma had been his regular correspondent, a constant companion throughout his rambling scientific journey. She knew and understood him, perhaps as well as anyone, and her death was a huge personal loss. 'I do not know what he would have done,' wrote Milly, 'but for his great niece, Eva

Biggs, who devoted herself to him as if she had been his daughter.' Galton also turned to Milly for comfort. 'We must contrive means of keeping in closer touch,' he wrote. 'Bessy and I write every week. You and I must do the same.'

Despite his personal problems, Galton's professional life was in extremely good shape. Sensing the reverse in public opinion about eugenics, he was quick to capitalise. In the autumn of 1904 he entered into negotiations with the University of London to establish a research fellowship in eugenics. He was happy to give £1,500 for a trial period of three years to cover salaries and expenses for the fellow and an assistant. A young zoologist called Edgar Schuster was drafted in from Oxford University and appointed as the first eugenics fellow, while University College provided rooms at 50 Gower Street in central London. There, in an imposing Georgian terraced house, the Eugenics Record Office was born.

Delighted by this latest development, Galton was keen to tell Milly all about it. 'You see now (1) that everything is done in the name of the University and (2) that the word "Eugenics" is officially recognised. I am very glad of all this as it gives a status to the Inquiry, so that people cannot now say it is only a private fad.' The *Pall Mall Gazette* agreed, congratulating the University of London and the Sociological Society for enabling Galton 'to develop and further promulgate his new study of *eugenics*'.

A letter from Germany confirmed that Galton's ideas were also catching on abroad. In 1904 a German eugenics journal, *Archiv für Rassen- und Gesellschaftesbiologie*, was founded by the anthropologist, Alfred Ploetz. A year later Ploetz created the German Society for Race Hygiene in Berlin. 'We take the highest interest in your eminent and important Eugenics,' he wrote to Galton, 'which is so closely connected with the subject of our *Archiv*, and shall keep our readers acquainted with the further development of your ideas.'

After so long in the wilderness, Galton was finally beginning to find an audience that was receptive to his ideas. His message of human improvement was no longer being ignored. Now, at last, he had a stage on which he could argue the logic of his vision.

It was while Galton was a student at Cambridge that the germ of his idea had first taken root. He had arrived at Trinity College as a bright and brainy pupil, but no matter how much effort and extra study he put into his work, he still couldn't compete with the best. To Galton it seemed that hard work alone was no match for innate talent. Delving into the issue a bit further he discovered that the most gifted students – the so-called senior classics and senior wranglers – often had close relatives who had themselves been the best in their respective years. Out of forty-one senior classics, for instance, six had a father, son, or brother who was also a senior classic. These statistics helped convince him that exceptional abilities can only be born, not made. His 1869 book, *Hereditary Genius*, was a more thorough and extensive development of these first, elementary impressions.

Cambridge may have got Galton thinking about hereditary inequalities, but it was *The Origin of Species* that provided his real awakening. As he consumed his cousin's book he felt his old theological baggage fall away, as his mind attuned itself to an entirely new kind of reasoning. Darwin had completely rewritten the history of life on earth. Now Galton began to think about rewriting its future.

If Darwin was right, then everything that had ever lived was bound together by one eternal struggle for food, for living space, and for mates. It was out of this never-ending battle for survival that the great diversity of life had sprung. Humans, aardvarks, cabbages, and coots: all were playing the same game. The only difference was that now we could comprehend the rules. Was it not churlish, Galton had asked, irresponsible even, not to use this knowledge to our advantage, and to follow the course that nature had laid out for us? If people could use selective breeding to improve the varieties of plants and animals, then why could they not do the same to themselves?

Not everyone, of course, could take part in the experiment. Active participation would be restricted to the lucky few. But this wasn't about the wants and needs of individuals. It was about the drive for national excellence. And it was every person's responsibility, from the most able to the most idiotic, to follow that faith. In eugenics, Galton saw the basis of an entirely new religion.

Before human improvements could be made, however, Galton realised the need for a better understanding of heredity. And while he never came up with an entirely satisfactory explanation of his own, he did manage to dish the dirt on Darwin's pet theory of pangenesis. The important implication of Galton's refutation of pangenesis was that qualities acquired through good nourishment and good education could never be inherited. It was, therefore, a waste of effort to try to improve a poor stock by careful nurturing, because any gains made during the current generation would have disappeared by the next. Far better, then, to breed from a superior stock to start with.

To gauge whether you are improving or not you have to know where you started from. So the first stage of Galton's campaign for eugenics was to take account of the nation's physical and psychological characteristics. Only then would it be possible to assess what effect a selective programme was having on the state of the nation. Inevitably, Galton's anthropometric investigations produced a bumper harvest of numbers and he had been forced to devise novel statistical techniques in order to make sense of it all.

Galton also worked hard at making his eugenic discriminations a little easier, hunting high and low for a physical index of mental character. Composite portraiture and fingerprints had shown initial promise but, ultimately, both had been found wanting. It was a disappointment, but no disaster, on the ultimate path to progress.

There had been other stumbling blocks along the way, not the least of which was the problem of regression. Galton's work on the statistics of human inheritance had, at one time, convinced him that eugenic selection on small variations would lead nowhere. Only human 'sports' – the J.S. Bachs of this world – could provide the impetus for change. But Karl Pearson had shown that Galton's biological interpretation of regression was wrong. Selection on small variations was not an impotent force. On the contrary, it could be the impetus behind a strategy of human improvement.

It was on the practical side of eugenics where criticism, in the past, had been most vehement. Eugenic marriage restrictions, it was claimed, were unrealistic and unworkable. Individuals would never accept a situation in which the state picked their partner; human passions would always get in the way. But marriage restriction was not a new idea, Galton argued. Human beings had grown used to living in cultures where partnerships were policed. Men in many African societies, for instance, had to demonstrate their physical strength before they were entitled to marry. In Austria and Germany it was illegal for paupers to marry. Even in Britain, marriage restrictions based on class and social status had been around for centuries. So why, Galton asked rhetorically, could laws not be extended to prohibit the marriage of parents destined to produce inferior children?

Although people might find the notion of eugenic marriage restrictions distasteful, Galton believed that they would be won over if a strong enough case was put before them. It was a giant moral leap to make, but not an insurmountable one. Thorough explorations of Africa and the landscape of his own mind had convinced him that religious and moral devotion was simply a response to cultural conditioning. The mind and its morals were essentially malleable and, with a bit of persuasion, could be shaped into any number of different forms. 'Any custom established by a powerful authority soon becomes looked upon as a duty, and before long as an axiom of conduct which is rarely questioned.' If eugenics was granted the status of a religion, then the sanctity of marriage would be preserved, not as a marriage made in heaven, but as one made in accordance with the natural law of heredity. In this new moral climate, he believed, a non-eugenic union would become as socially taboo as incest.

With the moral hurdles out of the way, all that remained were the details of the eugenic selection procedure. Here there was still lots to think about. The make-up of the selection panel, the form of the competitive examinations, the nature of the endowments, and the enforcement of restrictions: these were all things that would be ironed out in due course. But there were some simple and obvious things that could be done straight away that might bring about an immediate improvement. If all young married couples were divided up into three classes – desirables, passables, and undesirables – based on the probable civic worth of their offspring, then 'It would surely be advantageous to the country if social and moral support, as well as timely material help, were extended to the desirables, and not monopolised, as it is now apt to be, by the undesirables. Families which are likely to produce valuable citizens deserve at the very least the care that a gardener takes of plants of promise.'

Galton believed that the distribution of charity needed a complete overhaul. In a eugenic society it made no sense for the mentally and physically weak to command the lion's share of financial aid. It was simply a bad investment for the future. If there was money available then it should go to the healthy and the fit to support their winning hand. Money could come from public or private sources. Wealthy landlords, for instance, could be encouraged to sponsor impoverished men and women of outstanding biological merit, providing them with a healthy outdoor life in the country. The weak would not be forgotten, but attached to any charitable gifts they received would be a binding contract of celibacy. All undesirable life had to be drained from the human gene pool.

This was the outline of Galton's eugenic vision, his step-by-step guide to a better national breed. It was a daring and audacious plan, but Galton was somewhat realistic in his ambitions. He knew that none of it could happen overnight. While his lifetime's work had already won him converts to the cause, he realised that politicians and the wider public would require more convincing.

In fact the majority of the electorate seemed to be moving in exactly the opposite direction. In 1905 a Liberal government came to power, wielding a radical manifesto of its own. The Liberals tackled the poverty issue as a social malaise rather a biological one. The introduction of free school meals, old-age pensions, and national insurance were all welcome vindications of this optimistic outlook, and heralded the arrival of the welfare state. Of course the money for state reform had to come from somewhere. Direct taxation on the wealthy was the obvious solution, but it was a decision unlikely to find friends among the emerging eugenics crowd.

In the winter of 1905 Galton and Eva travelled down to the south-eastern coast of France, settling in, once again, to a luxurious life abroad. This time their vacation took in long stays at the more expensive hotels of Biarritz, Ascain, and St Jean de Luz. The change of scene gave Galton the chance to escape the cold, damp air of a London winter, and continue his scientific work in a climate more benevolent to his increasingly rheumatic body.

He was still pursuing an active scientific schedule, supervising the statistics flowing out of the newly formed Eugenics Record Office, and constantly tinkering with the fragments of old ideas. But compared to the frenetic pace of his prime, he had slowed down considerably. With so much more time on his hands Galton, almost grudgingly, began to rediscover the pleasures of reading.

Galton always clung eagerly to the literature of his youth. Greek myths, Shakespeare, and the poems of Tennyson and Byron were perennial favourites. As an adult he was never that fond of fiction. He seemed aloof from the imaginary world and its emotional unpredictability, as if it was all artificial nonsense that was some-how beneath him.

When he reached his eighties, however, Galton suddenly changed his tune, and started devouring novels as if making up for lost time. Books like Cervantes' *Don Quixote*, Jack London's *The Call of the Wild*, and Ernest Renan's *Antichrist* reflected his eclectic tastes. Yet his enjoyment still seemed to embarrass him. 'I have been occupying all my novel-reading hours with reading *Sir Charles*

Grandison,' he wrote to Milly, 'and am ashamed rather to say how much I am carried on with it.'

Early in the new year the French holiday was interrupted by sad news. On 7 January 1906 a telegram informed Galton of the death of his sister Elizabeth at the age of ninety-six. 'It is the last link with my own boyhood,' he explained to Milly, 'for Erasmus was at sea, etc., and knew little about me then. So much of interest to myself is now gone irrecoverably.'

With only his brother Erasmus remaining, Galton, now almost eighty-four, was beginning to feel hemmed in by death. His mind was still active, but he was becoming increasingly infirm, and the holiday in Biarritz was to be his last tour abroad. Although he made plans for an Italian vacation the following winter, his worsening rheumatism, asthma, and bronchitis meant that visiting the continent was no longer possible. When the French holiday came to a close in the spring of 1906, the travelling days of this great explorer were almost over. From now on, the English coast would be as far south as Galton would ever get.

London still held its attractions, so long as eugenics continued to thrive. The Eugenics Record Office on Gower Street was busy creating registers of noteworthy families, based on detailed studies of pedigrees. But Edgar Schuster, the office's research fellow, seemed reticent in his role. He had very little say over the office's direction. Galton was the man in charge, the one pulling all the strings. And besides, Schuster appeared more interested in the study of animals than the building of a better human race. After two years in the job he decided to call it quits.

Unsure of what to do next, Galton consulted his friend Karl Pearson. Galton had big plans for Pearson. Through talks with University College, Galton had set up a professorship to come into effect after his death. His will clearly stated his wish to see Karl Pearson appointed as University College's first Galton Professor of Eugenics. With this in mind, it made sense for Galton to relinquish his role and appoint Pearson as the new director of the Eugenics Record Office. Pearson agreed to the proposal on the condition that there could be some degree of collaboration between the office and his own Biometric Laboratory. The Eugenics Record Office was renamed the Galton Laboratory of National Eugenics, a replacement fellow was found for Schuster, and the Utopian think-tank was back in business.

One of the earliest and most influential studies to emerge from the revamped Galton Laboratory was an investigation into the demographics of various London districts. When David Heron, the new Galton Fellow, looked at fertility rates of different social groups within London, he found that the working classes were making a disproportionate contribution to each succeeding generation. The report did little to placate those already living in fear of national deterioration. The socialist reformer and Fabian Society member, Sidney Webb, blamed Irish and Jewish immigrants, together with 'casual labourers and the other denizens of the one-roomed tenements of our great cities'. Heron's report did little harm to Galton's growing eugenic cause.

In late 1907 Galton was approached by the publisher, Methuen, to write his autobiography. He agreed, and dashed off the book in under a year. In October 1908 *Memories of My Life* was released to generally good reviews and very good sales: the first edition of 750 copies sold out in a month. Galton's life contained so much of interest that his story guaranteed a good read. And yet, for all its popularity, it read like a book that had been written too fast. In *Memories*, Galton perfected his stream-of-consciousness style. With no concession to continuity, it was unpredictable and occasionally exciting stuff, and each new sentence offered a potential surprise. In one paragraph, for instance, a nostalgic reminiscence about the long Easter walks he took with his brothers-in-law suddenly broke off into a recipe for cheese sandwiches:

> Let me venture to describe my own views as to provisions suitable for a day's walk during a homely tramp. They are such as can be procured at any town however small, are tasty, easy to carry, exempt from butter, which is apt to leak out of

paper parcels, and are highly nutritious. They are two slices of bread half an inch thick, a slice of cheese of nearly the same thickness, and a handful of sultana raisins. The raisins supply what bread and cheese lack; they play the same part that cranberries do in pemmican, that nasty, and otherwise scarcely eatable food of Arctic travellers. The luncheon rations that I advocate are compact, and require nothing besides water to afford a satisfactory and sustaining midday meal. If sultanas cannot be got, common raisins will do; lumps of sugar make a substitute, but a very imperfect one.

Anyone hoping to find something of the inner Galton beneath the book's cover was sure to be disappointed. Galton's assessment of his life provided only a perfunctory analysis of the personality that had led it. This was very much a book about facts rather than feelings, and was, in many ways, more revealing for what it left out than what had been put in.

Galton did, at least, acknowledge that popular sentiment had been against him when he first aired his views on heredity and eugenics in the late 1860s. 'Now I see my way better,' he explained, 'and an appreciative audience is at last to be had, though it be small.'

As in most other cases of novel views, the wrong-headedness of objectors to Eugenics has been curious. The most common misrepresentations now are that its methods must be altogether those of compulsory unions, as in breeding animals. It is not so. I think that stern compulsion ought to be exerted to prevent the free propagation of the stock of those who are seriously afflicted by lunacy, feeble-mindedness, habitual criminality, and pauperism, but that is quite different from compulsory marriage. How to restrain ill-omened marriages is a question by itself, whether it should be effected by seclusion, or in other ways yet to be devised that are consistent with a humane and well-informed public opinion.

I cannot doubt that our democracy will ultimately refuse consent to that liberty of propagating children which is now allowed to the undesirable classes, but the populace has yet to be taught the true state of these things. A democracy cannot endure unless it be composed of able citizens; therefore it must in self-defence withstand the free introduction of degenerate stock.

It was just this kind of fighting talk that roused the growing Galton fan base. In 1907 a small group of enthusiasts, loosely affiliated to the Moral Education League, founded the Eugenics Education Society. The original Society was based in London, but branches soon sprang up in Manchester, Birmingham, Glasgow, and other British cities. Galton himself was instrumental in setting up the Society. He realised that eugenics needed a more organised and populist voice if it was ever to win over the public mind. But he also recognised that popularisation, if not done in a careful and controlled manner, could threaten the credibility of eugenics as a rigorous science. That concern was evident in an address he gave to the Eugenics Education Society in London in 1908. It was the last public lecture he ever gave, and it contained more than a hint of foreboding for the future.

He began on a cautious note. 'Experience shows that the safest course in a new undertaking is to proceed warily and tentatively towards the desired end, rather than freely and rashly along a predetermined route, however carefully it may have been elaborated on paper.' Already it was evident that certain members of the Society were, perhaps, a little too eager for their own good. Galton wanted foot soldiers to go out and spread the message, but he didn't want fanatics.

After his words of caution, Galton went on to describe the direction in which he hoped the Society would head. He wanted to build networks of local associations all over the country. Members should go out into their communities, giving lectures on eugenics and heredity, signing up new converts, and winning support from doctors, lawyers, and other prominent members of society. Each branch should also compile lists of notable individuals in their district. He wanted to know the names of the best people in each profession, as voted by their fellow workers, and their detailed family histories. And as all this information was being gathered, he wanted it communicated back to party headquarters in London, to be catalogued and digested.

Having outlined his ambitions for the Society, Galton returned to his opening theme, reiterating the need for prudence:

> A danger to which these societies will be liable arises from the inadequate knowledge joined to great zeal of some of the most active among their probable members. It may be said, without mincing words, with regard to much that has already been published, that the subject of eugenics is particularly attractive to 'cranks' . . . It cannot be too emphatically repeated that a great deal of careful statistical work has yet to be accomplished before the science of eugenics can make large advances.

Galton saw the Eugenics Education Society's role as complementary to the work being done by Karl Pearson at the Galton Laboratory in University College London. The Society's purpose was to popularise and build support for eugenics, while the Galton Laboratory laid the scientific foundation on which any future eugenic legislation would be based. But from the start there was friction between the two camps. The presence among the Society's members of risqué sexual reformers like Havelock Ellis was a big turn-off for Pearson. Yet Pearson was himself part of the problem. The Society needed news and information to go into its new publication, the *Eugenics Review* and, not unreasonably, looked to the Galton Laboratory for ideas. But Pearson's intransigence frustrated the Society's attempts at any kind of collaboration, and his determination to keep the work of the Galton Laboratory under wraps won him few friends within the Society. With no sign of compromise, the situation deteriorated rapidly. A slightly bemused Galton could only look on from the wings as each side became entrenched. As founder of the Laboratory and Honorary President of the Society, Galton hoped for a reconciliation. But he was now a frail old man, with little energy for confrontation.

Galton had become too weak to walk, and had to be carried or wheeled wherever he went. His infirmity effectively rendered him housebound, although he could still enjoy a trundle around Hyde Park or Kensington Gardens in his bath chair if the weather was fine. The death of his last remaining sibling, Erasmus, in February 1909, did little to stem the sense of his own imminent demise, and his letters from this period sometimes read like a journal of decay. Each week a new pain brought fresh evidence of deterioration, the cog turning one more notch towards his death. Yet despite his growing list of complaints, Galton seemed at ease with himself and his circumstances. 'The sunset of life is accompanied with pains and penalties, and is a cause of occasional inconvenience to friends. But for myself I find it to be on the whole a happy and peaceful time, on the condition of a frank submission to its many restrictions.'

If he needed any further acknowledgement that the establishment still admired him, it came with the announcement of his knighthood in the summer of 1909. Galton seemed embarrassed by the whole affair, imploring Eva not to 'make any fuss about it'. Two months later he greeted news of his appointment to the council of the German Eugenics Society with far more enthusiasm. 'The society . . . has five honorary members among whom are Haeckel and Weismann, and I am asked to be its honorary Vice-President, which honour I have gladly accepted. But I must work up my German!'

Eugenic events at home, however, were still being overshadowed by controversy. In 1910 the nit-picking between the Galton Laboratory and the Eugenics Education Society erupted into a full-blown public row. At the centre of the argument was a Galton Laboratory report into the physique and intelligence of the

Twilight years: Galton, aged eighty-seven, with his disciple Karl Pearson

children born to alcoholic parents. The investigation, summarised in *The Times*, found that parental alcoholism had no discernible impact on the health of the children.

The Eugenics Education Society contained a number of temperance reformers among its ranks, and the report went against all their hopes and expectations. For the chairman, Montague Crackenthorpe, it was the final straw. Worn down by Pearson's posturing, he used the letters page of *The Times* to launch a scathing attack on Pearson's entire biometric methodology. The issue of alcoholism, he asserted, lay completely beyond the bounds of biometry. Pearson was ill-equipped to answer the question, and his statistical assessments were effectively worthless.

Crackenthorpe had little understanding of Pearson's work, but then that was hardly the point. Ridiculing Pearson in public was all that mattered to him. With Pearson seething, Galton was forced to step in and publicly smooth things over, playing down rumours of rifts and rancour in the eugenics camp. Statistical results were always interesting, he insisted, but it should be left to the individual to make up their own mind about what those results actually meant.

In rubbishing Pearson's work the Eugenics Society was, in effect, criticising the work of their founder, Francis Galton. It was Galton, after all, who had helped to put biological statistics on the map. The Society's stance also suggested that it was ignoring the warnings Galton had given in his original address, and was racing ahead with its own independent agenda. Galton was not alone in his concerns. Eva Biggs was worried about her great-uncle's association with the unpredictable ambitions of the Eugenics Education Society. 'I don't like him, at the end of his life, being mixed up with such a set,' she wrote to Karl Pearson, 'and who knows that some day he may not be made answerable for their actions, for after all he invented Eugenics.'

But events were running away from Galton. For decades he had been the isolated, outspoken voice. Now he could only watch as the popularity of his ideas overtook him. In Germany, France, New Zealand, and the United States activists had followed his lead by creating eugenics societies and journals of their own. Across the world, eugenics movements were hailing him as their hero. 'It would please you to realise', wrote the American eugenicist, Charles Davenport, 'how universal is the recognition in this country of your position as the founder of the Science of Eugenics. And, I think, as the years go by, humanity will more and more appreciate its debt to you. In this country we have run "charity" mad. Now, a revulsion of feeling is coming about, and people are turning to your teaching.'

Galton sat out his eighty-ninth and final year in a wheelchair, smoking hashish to temper his worsening asthma. But he still had time and energy for one last piece of propaganda, a fantasy novel set in the eugenic state of 'Kantsaywhere'. The book was rejected by the publishers, Methuen, on the grounds of some absurdly unrealistic love scenes. Galton asked that the manuscript be destroyed after his death, and most of it was. But sufficient extracts remained to piece together an extraordinary tale. This was Galton's final take on his eugenic state, and the picture he painted was clear and unambiguous. Beneath the story's idle whimsy and join-the-dots plot, was a stark and totalitarian image of the future.

The book purports to be 'Extracts from the Journal of the late Professor I. Donoghue, revised and edited in accordance with his request by Sir Francis Galton, F.R.S.'. Donoghue, our hero, is an adventurer and a professor of vital statistics. One day, when he is out and about on his travels, he stumbles across the bustling, bucolic colony of Kantsaywhere. The 10,000-strong colony is governed by the trustees of a Eugenic College that represents the spiritual and social centre of the entire community. It is in the College that the young occupants of Kantsaywhere sit for the examinations that will determine their fate.

Donoghue immediately falls in love with the place and its people. He is awe-struck by the purity of character and purpose: 'They are a merry and high-spirited people, for whose superfluous energy song is a favourite outlet.' Their conversation contains little gossip or scandal, and their speech is refined and poetic. As Donoghue carefully notes, 'The 'arry and 'arriet class is wholly unknown in Kantsaywhere.'

Despite his obvious sympathies, Donoghue's arrival is treated with some suspicion. Like everyone else in Kantsaywhere, he must sit the examinations if he hopes to stay. The examination itself comes in two parts. The result of the provisional or poll examination determines whether you can continue to live and breed in Kantsaywhere. Those who fail this first test are either forcibly deported, or shipped off to a segregated labour camp for a life of hard graft and celibacy.

Donoghue, somewhat predictably, sails through his provisional exam. He is, nevertheless, impressed by its thoroughness, likening it to the kind of medical exam that a 'very careful Insurance Office might be expected to require'. Passing the poll entitles him to sit for the much more rigorous honours examination. Success here would entitle him to all the privileges and endowments that the Eugenic College could provide, including the delectable Miss Allfancy and her thoroughly 'mammalian' features.

The honours examination came in four parts: anthropometric; aesthetic and literary; medical; and ancestral. During the aesthetic and literary section Donoghue has to stand up and perform a song-and-dance routine in front of the examiners, read out some poetry, strike athletic poses, and display his marching skills. It was tough and exhausting work but Donoghue pulled it off. The high marks were his and, with them, a passport to a life of privileged procreation.

What happened next will never be known. That part of the book, sadly, never survived. But, based on what had gone before, it would be fair to assume that Donoghue sires some of the most capable human specimens ever to walk the earth. He is hailed as an example to us all, and his name becomes embossed above the door of every household in the community, a potent reminder of the ultimate possibilities of good breeding. Above all else, he would be remembered as a remarkable man who did much to change the nature of the human race.

In November 1910 the Royal Society seemed to assess Galton in similar terms when they awarded him the Copley Medal – their highest honour – in recognition of his life in science. Galton was delighted, but his joy was touched by a sense of sadness. He knew that the end was near, and fond memories came flooding in. 'People are always very kind to me, but I wish my Father and Emma were alive. It would have given them real pleasure.'

Galton retreated to a rented house in Surrey for the winter. But bronchitis and asthma had turned breathing into a burden that his fragile eighty-eight-year-old body could no longer sustain. He survived into the new year, still cheerful, but in great discomfort. On 9 January he wrote to Milly, 'I wish I had something interesting to tell you, but have nothing to say more, beyond affectionate wishes to you all, individually.' He was too weak to write any more. His prolific life was over. He lay in bed for a few more days, struggling for breath, drifting in and out of sleep. On the seventeenth he saw Eva come into the room. He heard the concern in her voice when she asked him how he was feeling. 'One must learn to suffer and not complain,' he replied. And moments later he was gone.

Epilogue

Birmingham's Forgotten Son

Birmingham's Tourist Information Office in the centre of town offers a tasteful selection of mugs and T-shirts, but blank looks on the subject of Francis Galton. The woman behind the counter is friendly and attentive but it's clear that she's never heard of him. She calls over to her assistant on the other side of the store. 'There's a fella here wants to know about Dalton, Francis Dalton.'

I've come to the city to look for Galton's childhood home, The Larches, the country manor where he spent the first eight years of his life. The house itself was knocked down long ago, swallowed up in Birmingham's inner-city expansion. But I'm still keen to investigate the original location, to see what's become of the spot where he was born.

Back inside the tourist office I try a fresh probe and ask about Joseph Priestley. Suddenly the mood changes. It's smiles all round for the father of oxygen chemistry. Priestley was one of Birmingham's great exports and everyone's heard of him. I ask if they know where he lived, but then it's back to the blank looks. I conclude that a trip to the local library is in order.

The central library must represent the apex of Birmingham's concrete philosophy, a monolithic shrine to sand and cement. The building looks naked and ugly, but the remarkably helpful staff more than make up for the architecture. A quick dip into the archives and within minutes I'm on my way, heading out into Birmingham's southern suburbs.

A couple of miles later I'm on the fringes of Sparkbrook, home to some of the best Balti in Britain. The area, however, appears to have gone through a bit of a slump since Galton's day. Now it seems not so much up-and-coming as down-and-going. I don't want to be too hard on Sparkbrook. But first impressions are not good. There are no trees. There are rubbish and graffiti everywhere. Retail outlets are few and far between. And those places that haven't been boarded up have become washing-machine repair shops, car breakers, and second-hand stores.

But the street names suggest that my search is getting warm. Erasmus Road honours Erasmus Darwin, Galton's grandfather and founding member of the Lunar Society. Priestley Road is next on the right. It's an auspicious name for what is, in all honesty, an awful street. Anyone who still believes that the 1970s were cool should come and have a look at the housing on Priestley Road. Sitting somewhat incongruously on the side of number ten is a plaque commemorating the site where Joseph Priestley did some of his most famous work. I don't know whether this is Birmingham City Council's idea of a joke, but the burnt-out car on the opposite side of the street provides a fitting memorial to Priestley and the powers of combustion.

For the next half hour or so I wander round the area looking for any clues to the past. A couple of hundred yards on from Priestley Road I find Larches Street, a cul-de-sac backing onto a sparse, grassy area with a children's playground. This has to be the spot. Had I been standing here 180 years ago I would have been inside the old Larches estate. Perhaps today's swings and roundabouts mark the place where Galton once played as a child. It's impossible to say. The entire landscape has been resurfaced. Even the grass looks new. Beyond the street name, there is absolutely nothing to remind you that Galton once lived here.

Galton's disappearance from Birmingham history is revealing. It's not as if the city doesn't recognise its scientific heritage. After all, Joseph Priestley gets a commemorative plaque, so why not Galton? Both men were scientific revolutionaries and social idealists. Yet human sympathies naturally side with Priestley. Galton, perhaps, is not the kind of individual to inspire widespread affection. He was far too extreme for that.

But peel away the less attractive aspects of his character, and there is still much to admire. Here was a man who taught himself how to survive in one of the most hostile places on earth. Namibia turned out to be the training ground on which his immense stamina, energy, and courage came to the fore. Of course, Africa also exposed a cruel and brutal side to his personality, but it was nothing compared to the sadism of some of his contemporaries then charging across the continent. For all his racial rhetoric, his discipline, and his bravado, Galton always took great pride in the fact that every man on his expedition, black and white, came back alive.

Galton was naturally suited to the hardships of African adventure. In Africa he seemed to find a freedom that he never found elsewhere, and it is interesting to think how his life would have developed if, instead of getting married, he had continued with a career in exploration. Perhaps he could have become another Livingstone, Shackleton, or Scott. But poor health hijacked any hope of that. So he set an entirely different course, heading off to explore the vast landscape of science.

Galton was the product of an age when anything seemed possible, when men and machines formed a seminal symbiosis, while God became an optional extra. Galton's entire scientific career is infused with the spirit of these optimistic times. Believing that measurement could be applied to anything and everything, he began exploring its unlimited possibilities.

He first illustrated his measurement potential with mapping, meteorology, and making a good cup of tea. But it was in his study of human faculty where he really made his mark. Heredity, anthropology, and psychology were all dragged out of the scientific shadows by Galton's measured approach. Questions of body and mind were undoubtedly elusive. But given the right perspective, he showed that they could be caught in the headlights of statistical investigation. In doing so, he put the study of human variety on a scientific footing, painting numerical pictures of the range of human difference among individuals, families, and populations.

For all his achievements, however, Galton never made one of those giant scientific leaps that distinguishes the revered from the merely remembered. Compared with Mendel or Darwin, for instance, his labours were less focused, more dispersed, and perhaps a little less memorable as a consequence. He was as much a facilitator as an originator, opening new scientific doors through which others would follow.

By trying to cram so much into his life, Galton was inevitably forced to cut corners, and there is a half-finished feel to much of his work. Having dipped his toe into one idea, he was always looking to move on to something else. Invariably, that something else was another untouched area of science. His apparent aversion to established scientific fields was, perhaps, his way of avoiding the atmosphere of competition that had so shattered his confidence as a young man at Cambridge.

Whatever his motivations, there is no doubting the enormous influence of his work. Galton's 1864 article, 'Hereditary Talent and Character', signalled the future direction not only of his own career, but of biology in general. Human genetics; intelligence testing; the debate about nature versus nurture; and eugenics all owe their origins to this one article. By placing the study of man in an evolutionary context, he was also anticipating the emergence of contemporary scientific realms like modern anthropology, sociobiology, and evolutionary psychology.

Galton's life perfectly encapsulates the conflicts and contradictions that can afflict scientific progress, and the moral ambiguities that can feed the birth of new ideas. Galton left us with some important scientific tools to deal with the modern world. But many of his greatest achievements were born out of his desire to build a foundation for eugenics. It is a peculiar, perhaps uncomfortable, thought that modern statistics and human genetics owe much to one man's misplaced faith in human improvement.

Whether Galton's hard-line stance on heredity was a consequence of his eugenic ambitions is impossible to say. Galton claimed that he was first alerted to striking instances of hereditary ability while he was a student at Cambridge, long before ideas of eugenics came into his head. Either way, the details hardly matter. His simultaneous introduction of hereditary determinism and eugenics made them appear as two sides of the same coin. In people's minds, if not in practice, the subjects have been associated with one another ever since.

Today, it is almost impossible to talk about heredity and human behaviour without arousing suspicion of hidden agendas. Since the Second World War, hereditary determinism, eugenics, and right-wing extremism have all been lumped together as part of the same package. But the full history of eugenics reveals a far more ambiguous picture. At the peak of its popularity in the 1920s and 1930s, the subject appealed to a broad range of political persuasions. Liberals, socialists, and even communists lined up alongside conservatives in support of eugenics policies.

It is ironic that it was not in Britain, but in that nation Galton considered 'strongly addicted to cant' – the United States – where eugenics first gained a foothold. In 1907 the state of Indiana introduced the world's first compulsory sterilisation law for so-called inferior types. Criminals, the mentally ill, and the insane were all lined up to be neutered. Another thirty states soon followed suit. The label of social degeneracy was a loose, umbrella term that gave ample scope for prejudiced interpretations. In some states the badge of inferiority was extended to include homosexuals and communists. By 1940 more than 35,000 US citizens had been sterilised in the name of social progress.

In Hitler's Germany it was the same, only worse. The extent to which Nazi ideology relied on any kind of scientific foundation to support its campaign of racial extermination is open to question.

But there is no doubt that Heinrich Himmler, Hitler's right-hand man, publicly embraced the theory of racial hygiene put forward by Galton's old admirer and correspondent, Alfred Ploetz, as he strove for the Ayran ideal.

Perhaps it is unfair to lay blame at Galton's door for the appalling tragedy of the Holocaust. But he was a significant contributor to what was a complex chain of events. Apart from his hereditary obsession, he was a moral relativist with a very weak faith in democracy. On occasion, his descriptions of eugenics sound uncannily prophetic of National Socialism.

The atrocities of the Second World War left biologists with little appetite for human genetics, and the subject went into serious decline. But genetics research didn't go away. It simply moved into areas not blighted by controversy. This sideways shift proved remarkably productive. The 1950s and 1960s saw a chain of landmark discoveries that delivered us into our modern, molecular age.

Molecular biology transformed the genetic landscape. It became routine to isolate individual genes, to 'grow' them in bulk, to store them in libraries, to sequence their code, and to shunt them around, from one organism to another. These new technologies were irresistible and, by the 1970s, had turned human genetics into a thriving field once again.

Throughout all these upheavals Galton's name endured. In 1904 he began an association with University College London that continues to this day. Galton's Eugenics Record Office at 50 Gower Street in Bloomsbury was the first human-genetics laboratory in the world. Today, its descendant lives on as the Galton Laboratory, now part of the Biology Department at University College London. There, his bust presides over daily seminars, witness to the latest news from the coal face of human genetics.

History, however, has exposed the limitations of Galton's vision. Today we accept that genes may influence many aspects of our behaviour. But subsequent work on nature and nurture has shown that the question demands a far more sophisticated perspective than Galton ever envisaged. Nature and nurture cannot be treated as separate, isolated entities: neither makes sense in the absence of the other, and they interact in complex and unpredictable ways.

Yet Galton's extreme view of human nature continues to affect our thinking. The success of the Human Genome Project has once again brought heredity to the fore, reopening old debates about biological reductionism and hereditary determinism. With the deluge of new genetic information, there is concern that we may be moving back towards a Galtonian, gene-centric view of the world, in which human variety and the problems of society can be conveniently and simplistically pinned down to our DNA, and in which prejudice and bigotry can flourish. Who can forget the infamous headline in the *Daily Mail*? 'Abortion hope after "gay genes" findings.'

In 1994 the publication of Richard Herrnstein's and Charles Murray's book *The Bell Curve: Intelligence and Class Structure in American Life* caused a sensation in the United States. The critics were divided. Those on the political left dismissed the authors as 'un-American' and 'pseudo-scientific racists'; critics on the right championed them as 'brave and respectable scholars'. Amid the book's 800-odd pages was the general thesis that variation in human intelligence and IQ could be explained largely by genes. Most controversially, Herrnstein and Murray pointed towards the disparity in the distributions of IQ between African–Americans and Caucasian Americans, and implied that genetics might also explain this difference.

It is more than a century since Galton added biology to the elusive recipe for human progress. We are still arguing about nature and nurture, the merits of public services and the welfare state; education, class, and hereditary privilege. As we look forward to the future, and a new technological century of genetics, all the indications suggest that the ghost of this witty, uncompromising, phlegmatic, wayward, and single-minded man will continue to haunt us.

A NOTE ON THE AUTHOR

In a previous life Martin Brookes was an evolutionary biologist in the Galton Laboratory at University College London.